# POLYSULFIDE OLIGOMER SEALANTS
## SEALANTS

Synthesis, Properties, and Applications

# POLYSULFIDE OLIGOMER SEALANTS

## Synthesis, Properties, and Applications

Yuri N. Khakimullin, DSc, Vladimir S. Minkin, DSc,
and Timur R. Deberdeev, DSc

APPLE ACADEMIC PRESS

Apple Academic Press Inc. | Apple Academic Press Inc.
3333 Mistwell Crescent | 9 Spinnaker Way
Oakville, ON L6L 0A2 | Waretown, NJ 08758
Canada | USA

©2015 by Apple Academic Press, Inc.

First issued in paperback 2021

*Exclusive worldwide distribution by CRC Press, a member of Taylor & Francis Group*
No claim to original U.S. Government works

ISBN 13: 978-1-77463-345-8 (pbk)
ISBN 13: 978-1-77188-029-9 (hbk)

### Library and Archives Canada Cataloguing in Publication

Khakimullin, Yuri N., author
Polysulfide oligomer sealants: synthesis, properties, and applications / Yuri N. Khakimullin, DSc, Vladimir S. Minkin, DSc, and Timur R. Deberdeev, DSc.

Includes bibliographical references and index.
ISBN 978-1-77188-029-9 (bound)
1. Oligomers. 2. Polysulphides. 3. Vulcanization. 4. Sealing compounds. I. Minkin, Vladimir S., author II. Deberdeev, Timur R., author III. Title.

QD382.O43P64 2015     547'.7     C2014-908461-7

### Library of Congress Cataloging-in-Publication Data

Khakimullin, Yuri N.
Polysulfide oligomer sealants: synthesis, properties, and applications / Yuri N. Khakimullin, DSc, Vladimir S. Minkin, DSc, and Timur R. Deberdeev, DSc.

pages cm
Translated from Russian edition published in 2007.
Includes bibliographical references and index.
ISBN 978-1-77188-029-9 (alk. paper)
1. Oligomers. 2. Sealing compounds. 3. Polymers. I. Minkin, Vladimir S. II. Deberdeev, Timur R. III. Title.

QD382.O43K4313 2015     620.1'99--dc23     2014049085

Apple Academic Press also publishes its books in a variety of electronic formats. Some content that appears in print may not be available in electronic format. For information about Apple Academic Press products, visit our website at **www.appleacademicpress.com** and the CRC Press website at **www.crcpress.com**

# ABOUT THE AUTHORS

**Yuri N. Khakimullin, DSc**

Yuri N. Khakimullin, DSc, is a Professor in the Department of Chemistry and Processing Technology of Elastomers at Kazan National Research Technological University in Kazan, Russia. He has written four monographs, has written over 150 articles published in scientific journals, and holds 31 patents. His research areas include the synthesis, technology and processing of rubber and sealants based on polysulfide oligomers and elastomers and processes of radiation degradation of polymers..

**Vladimir S. Minkin, DSc**

Vladimir S. Minkin, DSc, is a Professor in the Physics Department at Kazan National Research Technological University in Kazan, Russia. He is the author of four monographs, over 200 articles, and 12 patents. He has authored four monographs, written over 200 scientific articles, and holds 12 patents. His research interests include NMR in reactive oligomers; investigation of the structure and properties of sulfur-containing composite materials based on liquid thiokols; application methods of magnetic spectroscopy to study the structure, phase composition, molecular mobility of petroleum and petroleum products, and natural bitumen; and the development of new devices for rapid analysis and group composition heavy oil and natural bitumen.

**Timur R. Deberdeev, DSc**

Timur Deberdeev, DSc, is a Professsor in the Department of Processing Technology of Polymers and Composite Materials at Kazan National Research Technological University in Kazan, Russia. He is the author of two monographs, has written over 60 scientific papers, and holds 13 patents. His research areas include the synthesis, technology, and processing of rubber and polymers, PVC, and composite materials based on epoxyamine systems.

# CONTENTS

# LIST OF ABBREVIATIONS

| | |
|---|---|
| 1;1-NT; 2; 2-NT | liquid thiokol grades |
| 7-1; 7-20; TMGPh-11; TGM-3; MGPh-9 | grade of oligoesteracrylates |
| Agidol AF-2 | catalyst, a Mannich base Transamination Product |
| AM-05, AM-05K | Grades of Thiokol Sealants, Used in Construction Industry |
| BN70/30 | Petroleum Bitumen Grade |
| CED | Cohesion Energy Density |
| DDE | Disulfide-Disulfide Exchange |
| DG-100 | Channel Type Carbon Black Grade |
| DTA | Differential-Thermal Analysis |
| DTB | Density of Transversal Bonds |
| FTD | Functional Type Distribution parameters |
| HP-30; HP-470; HP-52 | Grades of Chloroparaffines |
| IPN | Interpenetrating Polymer Network |
| LT-1; LT-1K; SG-1; SG-1K | Grades of TPM-2-Based Sealants, Used in Construction Industry |
| MB | Mannich Bases |
| MNI ITEP | Moscow State and Design Institute of Typology and Experimental Design |
| MTD-2 | Natural Chalk Grade |
| MWD | Molecular Weight Distribution |
| NVB-2 | Liquid Thiokol Grade |
| OEDGA-50 | Oligoester based on Ethyleneglycol and Diethyleneglycol (1:1) Mixture and Adipinic Acid |
| OM-3 | Mannich Base |
| P-803; PM-50; PM-70 | Furnace Type Carbon Black Grades |
| PIC | Polyisocyanate |
| PN-1; PN-9119 | Grades of Unsaturated Polyesters |
| PSO | Polysulfide Oligomer Sealants |
| SFA | Synthetic Fatty Acids |
| SKUDF-2, FP-65 | Oligobutadienediol-Based Prepolymers with Isocyanate End Groups |

| | |
|---|---|
| SKUFE-4 | Polyester-Based Prepolymers with Isocyanate End Groups |
| SKUPFL-100 | Oligooxymethyleneglycol-Based Prepolymer with Isocyanate End Groups and Molecular Weight of 1000 |
| SKUPPL-4503; | Oligo Oxypropyleneglycol-Based Prepolymers |
| SKUPPL-5003 | with Isocyanate End Groups and Molecular Weight of 4503 and 5003 Correspondingly |
| Solid thiokol "DA" | Solid Thiokol Grade |
| STIZ-30 | Thiokol Sealant |
| Thiuram D | Tetramethylthiuramdisulfide |
| TMA | Thermomechanical Analysis |
| TMD | Transversal Magnetization Decay |
| TPE | Thiol-Polysulfide Exchange |
| TPM-2 | Thiol-Containing Polyester with End SH-Groups |
| U-30M; U-30MES-5; U-30MES-10; UT-31; UT-32; UT-34; 51UT-36; 51UT-37; 51UT-38; VIT | Grades of Thiokol Sealants |
| U30 E-10; U30 E- 5 | Grades of Sealing Pastes |
| USPE | unsaturated polyesters |
| E-40, ED-20 | Epoxy Diane Resin Grades |

# LIST OF SYMBOLS

| | |
|---|---|
| $\xi$ | adhesion parameter |
| $\nu_{chem}$ | chemical density of transversal bonds |
| $\nu_{eff}$ | effective chain density |
| $\delta p$ | solubility parameter |
| $k_e$ | rate constant |
| $lg\eta$ | dependence of viscosity |
| $lg\gamma$ | shear rate |
| $H_2^2$ | second moment values |
| $[MnO_2]cryst$ | manganese oxide crystallite |
| $R'$ | alkylene |
| $R'_{in}$ | radical products of Mannich base decay |
| $R_{in}^*$ | radicals |
| $T_{21}(t)$ | induction period |
| $T_{2final}$ | final relaxation times |
| $T_{flow}$ | viscous-flow transition temperature |
| $T_{glass}$ | glass-transition temperature |
| $T_{soft}$ | softening temperature |
| $\bar{f}_n$ | number-average functionality |
| $\gamma$ | shear rate |
| $\varepsilon_{rel}$ | relative elongation |

# ABSTRACT

This book deals with problems of synthesis, vulcanization, modification, and the study of structure and properties of highly filled sealants based on polysulfide oligomers.

The book summarizes data on chemistry and synthesis technology of liquid thiokols and thiol-containing polyesters, and their structure and properties. It provides scientific information on chemistry and mechanisms of liquid thiokols vulcanization by oxidants and in polyaddition reaction. Basic formulation principles for sealing compositions are given; their properties and application range are described.

The monograph provides the results of authors' research on vulcanization and modification of thiokol sealants, using thiokol-epoxy resin copolymers, unsaturated polyesters and isocyanate prepolymers of a various nature. It combines research on vulcanization mechanisms of polysulfide oligomers by manganese dioxide, sodium bichromate and zinc oxide, as well as on the structure and properties of sealants, based on liquid thiokol and TPM-2 polymer, depending on the nature and ratio of used oligomers. This publication also gives information on the influence of fillers on vulcanization kinetics, rheological and physical-mechanical properties of sealants, depending on the nature of PSO.

The book is meant for scientific and engineering personnel at institutes, science centers and companies pursuing research in design, structure and properties of reactive oligomers and related compositions and dealing with their production and application, as well as for professors, postgraduates and graduate students in the field of chemical technology education.

# PREFACE

Polysulfide oligomers are reactive low-molecular rubbers and the representatives of the earliest feedstock for industrial production of cold cure sealants in the world. The main reason of such sealants being widely used in modern construction, mechanical engineering and aircraft industries relates to their unique range of properties and corresponding excellent performance. This range combines outstanding gas impermeability and atmosphere resistance with oil and gasoline resistance. In addition, high demand for polysulfide oligomers roots in their effective curing ability via oxidation of end SH-groups as well as in their high reactivity to various functional groups, such as epoxy, acrylate and isocyanate ones. It is the basis for effective chemical modification of sealants and control of product composition and related properties of at the stage of sealing blend preparation. On the other hand, sealants are commercially attractive because of their high stability during pre-application storing. All the aforesaid proves the technology of curing via end SH-groups to be extremely attractive and perspective.

This monograph is dedicated to aspects of synthesis, properties and application of polysulfide oligomer sealants. First two chapters provide a survey of existing technologies for synthesis of liquid thiokols and oligomers with end SH-groups (mainly using oligooxypropyleneglycols and thiol-containing polyesters) as well as curing methods, composition and application of polysulfide oligomer sealants. These chapters also provide systematized up-to-date information, available in literature after Lucke's monograph (Lucke H. Aliphatic Polysulfides. Monograph of an elastomer) publications by Huthig and Wepf in 1994. Next chapters summarize research results of monograph authors in curing, chemical modification and filling of sealants, based on liquid thiokol and TPM-2 polymer, as well as the study of their structure and performance (including service properties characterization) at a high filling rate.

The book may be useful for researchers with area of interest covering synthesis and application of polysulfide oligomers as well as for various scientists and engineers engaged in synthesis, modification and processing of reactive oligomers.

Prof. Gennady E. Zaikov, Editor
Head of Polymer Division,
N.M. Emanuel Institute of Biochemical Physics,
Russian Academy of Sciences

# INTRODUCTION

There is a stably growing modern tendency of consumption polymer composites designed for outdoor application. Composite materials containing unsaturated elastomers and oligomers are the most popular components for sealing and waterproofing materials as they demonstrate durable and effective resistance to such aggressive factors as ultraviolet radiation, ozone and water in a broad temperature range of –60 to +100°C.

Sealing materials based on reactive oligomers become more and more important today. Polysulfide oligomer sealants (PSOs), with thiokol as the main representative, occupy a special place among these materials. However, their application in Russia was limited to special purposes until the 1980s. These products used to take only 10% of construction, market, while produced thiokol sealants were low-filled.

The consumption pattern of thiokol sealants is quite different today, as they have become highly demanded for sealing interpanel seams and especially for sealing glass packets. Therefore, considering the growing world deficit of liquid thiokols, there is a necessity of application of highly filled PSO-based compositions. The scientific basis for vulcanization and modification of thiokol sealants has been mainly developed in studies of N. P Apuhtina, R. A. Shljahter, L. A. Averko-Antonovich, P. A. Kirpichnikov, R. A. Smyslova, E. M. Fettes, S. Iorzhak Dzh, M. Berenbaum, E. Dahsel't, T. S. P. Li, G. Ljuke et al. However, development of highly filled sealants based on liquid thiokol or thiol-containing polyester–TPM-2 polymer has revealed almost total deficiency of research on the influence of inert fillers on their vulcanization process and physical-mechanical and service properties. It should be noted that research, which acquires a special interest today, concerns introduction of more available reactive oligomers and oligomers of another nature into thiokol-based sealing formulations, and the study kinetic behavior of vulcanization network formation processes, modification and vulcanization mechanisms. The results of aforementioned research provide a scientific ground for modification and vulcanization processes, the structure of forming vulcanization networks, and the properties of PSO-based sealants in the highly filled state.

This book is a logical continuation of earlier publications by Professor L. A. Averko-Antonovich, with coauthors [1], Professor V. S. Minkin [2] and the authors [3]. In view of the aforesaid and considering the fact of the absence of recent summarizing publications in this area, this monograph provides data on the chemistry and technology of PSO production, on vulcanization, modification and application of related sealants, as well as the results of research, pursued by the authors themselves.

## KEYWORDS

- chemistry and synthesis technology
- cold cure sealants
- industrial production
- liquid thiokols and thiol-containing polyesters
- oligomers
- polysulfide
- reactive low-molecular rubbers

## REFERENCES

1. Averko-Antonovich, L. A., Kirpichnikov, P. A., & Smyslova, R. A. (1983). Polysulfide Oligomers and Related Sealants. Leningrad: Himija, 128 p.
2. Minkin, V. S. (1977). NMR in industrial polysulfide oligomers (in Russian). Kazan, "ABAK," 222 p.
3. Minkin, V. S., Deberdeev, R. Ja., Paljutin, F. M., & Khakimullin, Yu. N. (2004). Industrial Polysulfide Oligomers: Synthesis, Vulcanization, Modification (in Russian), Kazan, Novoe znanie, 175 p.

**CHAPTER 1**

# SYNTHESIS, STRUCTURE AND PROPERTIES OF POLYSULFIDE OLIGOMERS

## CONTENTS

## 1.1   CHEMISTRY AND PRODUCTION TECHNOLOGY OF POLYSULFIDE OLIGOMERS

Aliphatic polysulfides (or thiokols) are oligomers having fragments with disulfide bond and containing two or more mercaptan groups: HS − R (SS–R$'$)$_n$– SH. The term "Thiokol" originates in the trademark of polysulfide oligomers produced by the "Thiokol Chemical Corp." (USA) (later known as "Morton Inter. Inc.") [1].

Polysulfide oligomers (PSO) are reactive oligomeric compounds forming, if cured, sealants with a unique range of properties. High thermodynamic flexibility and the presence of chemically bound sulfur in the main chain of such compounds provide the respective sealants with a high fuel resistance, gas-tightness, water resistance as well as increased resistance to ultraviolet, solar radiation and ozone thanks to the main chain's saturation. PSO-based sealants have the ability of noncontractive low-temperature curing and are also highly durable. Acetal groups, on the other hand, favor acidic and basic molecular hydrolysis, but the small additives of a branching agent, introduced during a synthesis of oligomer, increase resistance of PSO vulcanizate to aggressive media both at normal and increased temperatures.

The first commercial production of liquid Thiokol based on 2, 2¢-dichlorodiethylformal with a common name "LP" (LP-2 × LP-3) was deployed in the USA by Thiokol Chemical Corp. in 1943 involving the technology proposed by Patrick and Fergusson [1]. Sodium disulfide was used there as a representative of sodium polysulfide. The production of liquid Thiokol of sodium tetra sulfide was deployed in USSR in 1959. The technology worked out by professor Apouhtina and colleagues was used there [9, 10].

Liquid thiokols are made today of high-molecular polysulfide through their reductive decomposition [9, 10]. Therefore, their production cycle consists of two main stages: the synthesis of a high molecular weight polysulfide and its further ditri or tetrasulfide bond decomposition, resulting in the reduction of a molecular weight (MW) of the produced super polymer (usually down to $M_p$ » 1000,7000).

The synthesis of liquid Thiokol is based on the polycondensation reaction of di- or tri halogen derivatives of organic compounds with sodium di- or tetrasulfide.

The most widespread monomer is 2, 2¢-dichlorodiethyl formal, which provides the highest thermodynamic flexibility for macromolecular chains [2, 11]. 2, 2¢-dichlorodiethyl formal is industrially produced by the acid-catalyzed interaction of ethylene chlorohydrine with formaldehyde in the presence of such compounds as dichloroethane, which are able to extract water zoetrope out of the reaction zone.

$$2\ HOCH_2CH_2Cl\ +\ CH_2O\ \longrightarrow\ ClCH_2CH_2OCH_2OCH_2CH_2Cl\ +\ H_2O$$

There is also an opportunity of the direct 2, 2¢-dichlorodiethylformal syntheses of ethylene oxide, hydrogen chloride and formaldehyde [1, 12].

The method of industrial production of sodium polysulfide is the interaction of caustic soda (40%) with sulfur.

$$6\ NaOH\ +\ (2n+1)S \longrightarrow\ 2\ Na_2S_n\ +\ Na_2SO_3\ +\ 3\ H_2O$$

The synthesis is carried out in a vertical agitated reactor at the temperature of 90–95°C. Sodium sulfite, formed during the reaction, interacts with sulfur and forms sodium thiosulfite. Sodium thiosulfate is a byproduct, which is removed in a wastewater during the synthesis. Another way to obtain sodium polysulfide is the reaction involving sodium sulfide:

$$Na_2S\ +\ (n-1)S\ \longrightarrow\ Na_2S_n$$

The reaction also takes place in water at the temperature of 90–95°C, but byproducts are not formed in this case.

The second method is more preferable due to reduced amount of forming waste. In is up to a producer to choose the method of sodium polysulfide synthesis, taking raw materials availability and economy into account.

Introduction of 0.1÷2.0% mol. of three-functional-1,2,3-trichloropropene (TCP) together with bifunctional monomers gives an opportunity to synthesize branched oligomers, whose vulcanizates are not subjected to a distinct cold-temperature flow and possess better physico-mechanical characteristics, than vulcanizates of linear oligomers.

The majorities of industrial-grade liquid thiokols have a weakly branched structure and contain HS end groups.

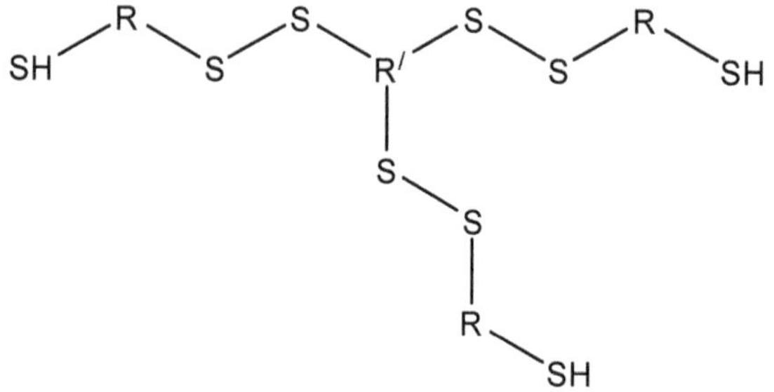

$$R = - CH_2 -CH_2 -O -CH_2 -O -CH_2 -CH_2 -$$
$$R' = - CH_2 -CH -CH_2 -$$
$$|$$

Liquid Thiokol is obtained here in the interfacial liquid polycondensation of sodium polysulfites with organic halogens being mainly aliphatic. The mechanism of 2,2$^\varepsilon$-dichlorodiethylformal with sodium tetrasulfide reaction is shown below:

$$nClCH_2CH_2OCH_2OCH_2CH_2Cl + nNa_2S_4 \longrightarrow$$

$$(-CH_2CH_2OCH_2OCH_2CH_2 -SSSS-)_n + 2 nNaCl$$

The reaction mechanism is the nucleophylic substitution of chlorine groups by a polysulfide-anion in water dispersion at the temperature of 80–100°C and with exothermic heat rejection (450 cal. per 1 kg of 2,2$^\varepsilon$-dichlorodiethylformal). The dispersing agent in a polycondensation process is magnesium hydroxide made of magnesium chloride.

The chain grows in the following reactions [13]:

$$Cl-R-Cl + Na_2S_n \longrightarrow Cl-R-S_n-Na + NaCl$$
$$Na-S_n-R-Cl + Na-S_n-R \longrightarrow \sim S_n-R-S_n-Na + NaCl$$

$$\sim S_n-R-S_n-Na + Cl-R-S_n\sim \longrightarrow \sim S_n-R-S_n-R-S_n\sim + NaCl$$

Polycondensation is usually carried out at the excess (up to 30%) of sodium polysulfide that provides formation of a polymer with –SNa end groups, their interaction results in increase of polymer's molecular weight:

$$\sim S_n\text{-}R\text{-}S_n\text{-}Na + NaS_n\text{-}R\sim \longrightarrow \sim S_n\text{-}R\text{-}S_n\text{-}R\sim + Na_2S_n$$

If a tetrasulfide polymer is used, dispersion is desulfurized by caustic soda, sodium sulfite, sodium hydrosulfide or sodium sulfide:

$$\sim(CH_2CH_2OCH_2OCH_2CH_2\text{-}SSSS)_n\sim \xrightarrow{NaOH} \sim(CH_2CH_2OCH_2OCH_2CH_2\text{-}SS)_n\sim$$

Excess of sodium polysulfide and alkali in a desulfurized dispersion is washed out by water and then sodium hydrosulfide is added in the presence of sodium sulfite, resulting in splitting of disulfide bonds in nucleophylic substitution reaction:

$$\sim R\text{-}S\text{-}S\text{-}R\sim + Na\text{-}HS \longrightarrow \sim R\text{-}SNa + H\text{-}S\text{-}S\text{-}R\sim$$

$$\sim R\text{-}S\text{-}SH \longrightarrow \sim R\text{-}SH + S$$

$$S + Na_2SO_3 \longrightarrow Na_2S_2O_3{}^+$$

$$\sim R\text{-}S\text{-}SH + Na_2SO_3 \longrightarrow \sim R\text{-}SH + Na_2S_2O_3$$

The molecular weight of forming liquid thiokol is regulated by the quantity of sodium hydrosulfide and its ratio to sodium sulfite.

Water dispersed liquid thiokol is coagulated by an acid. Mercaptide end groups are hereby transforming into mercaptan groups and the surface magnesium hydroxide is simultaneously destroyed:

$$2\sim R\text{-}SNa + H_2SO_4 \longrightarrow 2\sim R\text{-}SH + Na_2SO_4$$

Oligomer is further washed out from an acid and salts, dried, mixed with other batches of thiokol, which are then homogenized due to corresponding thiol-disulfide and disulfide-disulfide exchange.

The principal production process flow sheet [1] with some additions is shown in Fig. 1.1, where 1 is the reactor for sodium polysulfide synthesis; 2 is the reactors for synthesis of formal; 3 is the rectifying column, 4 is the mixer for formal and TCP; 5 is the mixer for dissolution of a crystalline sodium chloride; 6 is the reactor for the synthesis of a high molecular weight thiokol and its further desulfurizing, if sodium tetrasulfide is used; 7 is the tank (mixer) where the latex dispersion is washed out; 8 is the reactor for splitting a dispersion of high molecular weight polymer; 9 is the washing and coagulation mixer; 10 is the mixer for raw liquid thiokol; 11 is the accumulating tanks for liquid thiokol; 12 is the homogenizing mixer for liquid thiokol batches; F is the filter; D is the drier; and C is the centrifuge.

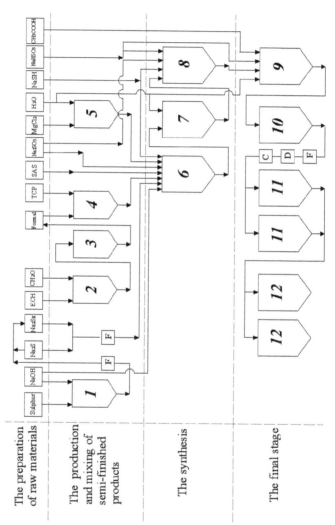

**FIGURE 1.1**    The principal production process of liquid Thiokol.

Reactors, tanks and pipelines are made of stainless acid and alkali resistant steel, with properties of raw materials and semi-finished products taken into account. Standard aluminum-protected (anodized) steel can be

also used, as well as acid and alkali resistant glass enamel. The process of liquid thiokol production can be conventionally divided into four steps [1]:
1. the preparation of raw materials;
2. the production and mixing of semi-finished products;
3. the synthesis; and
4. the final stage, combining liquid thiokol extraction, drying and homogenization. The first step is the one where raw materials are loaded and dosed; the second step is the synthesis and filtration of sodium polysulfide (1), the synthesis of formal (2) and its rectification (3), preparation of formal charging (4) and making of 25% water solution of magnesium chloride (5). The third step is the turn for equilibrium polycondensation and desulfurizing of a high-molecular weight Thiokol in dispersion (6), washing of dispersion for alkaline drainage disposal (7) and splitting of macromolecules (8). At the fourth step, (9) liquid thiokol dispersion is washed and coagulated. Dry liquid thiokol is pushed into mixing accumulator tank (10) and further fed to centrifuging, dried in vacuum and filtered. Liquid thiokol, when dried and filtered, is pushed into storage tanks (11). Mixers (12) homogenize various thiokol batches at 70°C. Then thiokol comes through categorization procedure and poured into containers to be delivered to a customer.

Major disadvantages of the industrial technology are its multistage structure and insufficient selectivity of polycondensation and splitting, Therefore, macromolecules with functionality defects can form and polymer branching grows considerably [13, 14], that has the extremely adverse effect on the environment. 4–5 tons of wastes and including 60 $m^3$ of wastewater form per 1 ton of produced liquid thiokol.

There were continuous attempts to design the low-waste technology of PSO production recently. Various solutions have been proposed, which are based on substitution of chlorine atoms in aliphatic polychloroderivatives to SH-groups with further oxidation of formed polythiols by atmosphere oxygen or sulfur to disulfide oligomers with the required molecular weight [15, 16]. Produced polymers have low molecular weight and functionality and the related cured compositions have unsatisfactory physico-mechanical properties. A good selectivity is demonstrated by a one-stage production of liquid thiokols via catalyzed oxidation of dimercaptan by compounds of iron and chrome [17]. This method is more practically feasible, but conversion of end CI-groups does not exceed 90% at the dimercaptan formation reaction. It impedes growth of a polymer chain and formation of a polymer with required functionality during polymerization. The following method of liquid thiokol production seems to be more attractive: interaction of a mixture (batch) of 2,2$^\ell$-dichlorodiethylformal and dichloropropane with a combination of sodium hydrosulfide and complex mixture of sodium hydrosulfide and sulfur at the concentration of 0.1–4.0 mol and 2–3 mol per 1 mol of batch correspondingly [18]. The process is single-stage. The charging is gradually added to the mixture of sodium hydrosulfide and sodium hydrosulfide-sulfur complex during 1,5–2 h at 70–80°C and then synthesis is carried out at the temperature of 80–98°C during 5–9 h. The reaction blend is cooled down to 30–40°C at the end of the synthesis, then mixed with toluene at the ratio of 2.5–3.0 parts of toluene per 1 part of a polymer. The water phase is further separated from the organic one. The organic phase is washed by water and dried. High reactivity of sodium hydrosulfide-sulfur complex and absence of side reactions lead to full substitution of end chlorine groups to SH-groups, giving an opportunity to obtain oligomers with controlled molecular weight and 97–98% yield, having almost no end chlorine groups (0.01–0.05%). The aforementioned technology can be implemented in a form of continuous process flow sheet with a recycle and minimum amount of waste.

Another way to produce PSO is the polycondensation of Bunte salts with sodium hydrosulfide in water and at the presence of toluene [19–22]. This method attracts by its the opportunity to carry out both synthesis and separation of liquid thiokol in one reactor with a minimum amount of waste (700 kg of NaCl per 1 ton of liquid thiokol) and to produce highly pure and tailored thiokols. The method of liquid thiokol production using Bunte salts is described below [23]:

$$\text{Cl-R-Cl} + 2\,\text{Na}_2\text{S}_2\text{O}_3 \longrightarrow \text{R(S-SO}_3\text{Na)}_2 + 2\,\text{NaCl} \qquad (1)$$

$$(n+1)\,\text{R(S-SO}_3\text{Na)}_2 + (1{,}5n+2)\text{NaSH} \longrightarrow$$

$$\text{HSR(SSR)}_n\text{SH} + (1{,}75n+2)\text{Na}_2\text{S}_2\text{O}_3 + (0{,}75n)\text{H}_2\text{O} \qquad (2)$$

An unusual stoichiometry of the process can be explained by simultaneous occurring of several reactions:

$$\sim\text{R-SSO}_3\text{Na} + \text{NaSH} \longrightarrow \sim\text{R-SSH} + \text{Na}_2\text{SO}_3 \qquad (3)$$

$$\sim\text{R-SSH} + \sim\text{R-SSO}_3\text{Na} \longrightarrow \sim\text{R-S-S-S-R}\sim + \text{NaHSO}_3 \qquad (4)$$

$$\sim\text{R-SSH} + \text{Na}_2\text{SO}_3 \longrightarrow \sim\text{R-SH} + \text{Na}_2\text{S}_2\text{O}_3 \qquad (5)$$

$$\sim\text{R-S-S-S-R}\sim + \text{Na}_2\text{SO}_3 \longrightarrow \sim\text{R-S-S-R}\sim + \text{Na}_2\text{S}_2\text{O}_3 \qquad (6)$$

$$2\,\text{NaSH} + 4\,\text{NaHSO}_3 \longrightarrow 3\,\text{Na}_2\text{S}_2\text{O}_3 + 3\,\text{H}_2\text{O} \qquad (7)$$

When the Eq. (2) is carried out, the order introduced reactants so as the duration and temperature of their mixing are very important. Liquid thiokol with a molecular weight from 2000 to 4000 is produced by the addition of sodium hydrosulfide solution to Bunte salt solution. Changing of mixing temperature within 20–70°C makes it possible to obtain liquid thiokols with tailored content of mercaptan groups as well as of molecular weight and viscosity. The synthesis is carried out in the presence of toluene. The properties of liquid thiokols depending on their synthesis temperature are given in Table 1.1.

**TABLE 1.1**   The Influence of a Mixing Temperature on the Properties of PSOs and their Vulcanizates*

| Parameter | Mixing Temperature, °C | | | |
|---|---|---|---|---|
|  | 20 | 30 | 40 | 50 |
| Viscosity at 25°C, Pa·s | 40 | 38 | 35 | 30 |
| The content of SH-groups, % | 2.1 | 2.2 | 2.3 | 2.5 |
| Shore hardness A, conv. Units | 67 | 66 | 65 | 65 |
| Conventional tensile length, MPa | 3.3 | 3.2 | 3.2 | 3.2 |
| Relative elongation at rupture, % | 240 | 260 | 250 | 240 |
| Relative residual elongation after rupture, % | 4 | 4 | 4 | 4 |

There are methods of liquid thiokol production using alternative dihalogenides. Oligomers prepared by partial or entire substitution of 2,2¢-dichlorodiethylformal: (1) to ($\text{ClCH}_2\text{CH}_2\text{-O-CH}_2\text{CH}_2\text{-O-CH}_2\text{CH}_2\text{-Cl}$); (2) have been proposed. It results in formation of liquid thiokol with increased polarity and vapor permeability and Therefore, promotes the rate of curing of sealants by atmospheric moisture and ($\text{Cl-(CH}_2)_{10}\text{-Cl}$); (3) leading to formation of liquid thiokol with reduced vapor permeability (Table 1.2).

The change of 20% mol. of 2,2¢-dichlorodiethylformal to a,a¢- dichloro-p-xylene; (4) also reduces the throughput rate of steam [6, 24].

**TABLE 1.2**   The Properties of Modified Liquid Thiokols

| Oligomer structure | Vapor permeability, g/ $(m^3 \cdot day)$ | Permeability, Conv. Units | $T_{10}, °C*$ |
|---|---|---|---|
| 1 | 20 | 0,0761 | 270–290 |
| 2 | - | 0,0960 | 289 |
| 3 | 11,25 | 0,0437 | 314 |
| 4 | 13,5 | 0,052 | 293 |

*The temperature corresponding to a 10% mass loss at the 20°C/min heating rate (by the thermo gravimetric analysis).

Bis-4-chlorobutylformal and bis-4-chlorobutyl ether [5] are used to increase frost resistance.

a,a¢-dichloro-n-xylene and bis-chloromethylthiofene or their mixtures with dichloroethane [5, 6, 25] (Table 1.3) are used in poly condensation process to increase strength, heat and chemical resistance of sealants based on liquid Thiokol. If bis-3-chloro-3-oxypropyl ether ($ClCH_2CH(OH)$ $CH_2OCH_2CH_2OCH_2CH(OH)CH_2$-Cl) is used as a dihalogenide, liquid thiokol with OH-groups in its side chain can be synthesized, and Therefore, related sealants can be highly adhesive to various substrates [26].

**TABLE 1.3**   Chemical Resistance and Physico-Mechanical Properties of Sealants based on Copolymer Thiokols

| Thiokol | Swelling, %, 29 days | | | | Conventional strength at rupture, MPa | Relative elongation, % | Shore hardness A |
|---|---|---|---|---|---|---|---|
| | Benzene | Toluene | $CCl_4$ | Acetone | | | |
| Dichloroethane | 43.4 | 30 | 28 | 2.8 | – | – | – |
| Dichlorodiethylformal | – | – | – | – | 6.7 | 162 | 88 |
| Dichloroethane with β-chloroethylformal | 98 | 55 | 31 | 28 | – | – | – |

**TABLE 1.3** *(Continued)*

| | | | | | | | |
|---|---|---|---|---|---|---|---|
| Dichloroethane with bis-chloromethylthiophenol | | | | | | | |
| 90:10 | 15 | 18 | 8 | 1.7 | 7.3 | 75 | 90 |
| 80:20 | 11.2 | 8 | 5 | 1.3 | 7.5 | 50 | 92 |
| 70:30 | 12.2 | 6 | 3 | 0.9 | 7.9 | 50 | 91 |
| Dichloroethane with α, α`-dichloro-n-xylene | | | | | | | |
| 90:10 | – | – | – | – | 7.2 | 160 | 94 |
| 80:20 | – | – | – | – | 7.8 | 110 | 95 |
| 70:30 | – | – | – | – | 8.2 | 75 | 91 |

There is thiokol marked ZL-560 and based on 2,2¢-dichlorideethyl formal with increased content (50%) of sulfur and having the following general formula [27, 28]:

$$HO\text{-}CH_2\text{-}\underset{\underset{S}{\|}}{S}\text{-}(R\text{-}S_{3,5})_{3\text{-}30}\text{-}R\text{-}S\text{-}CH_2\text{-}OH$$

where $R=(\text{-}CH_2CH_2\text{-}OCH_2\text{-}OCH_2CH_2\text{-})$

End methyl groups, which stabilize such oligomers during storage or at contact with water, are supposed to decompose forming –SH end groups in the following reaction:

$$\text{-}RSCH_2OH + H_2O \longrightarrow \text{-}RSH + CH_2O + H_2O$$

Related sealants possess minimum moisture permeability and increased resistance to oil and-gasoline.

An alternative scheme of synthesis of polysulfide oligomers, based on dithiodiethyleneglycol with hydroxide end groups, is proposed, which is similar to that for liquid thiokol [28]. The synthesis includes two steps:

1. dithiodiethyleneglycol synthesis

$$2HOCH_2CH_2Cl + Na_2S_2 \longrightarrow HOCH_2CH_2SSCH_2CH_2OH + 2\ NaCl$$

2. polyformal synthesis

$$nHO(CH_2)_2 S_2 (CH_2)_2 OH + (n-1)CH_2O \longrightarrow$$

$$HO[(CH_2)_2 S_2 (CH_2)_2OCH_2O]_{n-1}(CH_2)_2 S_2 (CH_2)_2OH + (n-1)H_2O$$

The production of such oligomers leads to a considerably reduced amount of wastes comparing production of a liquid Thiokol. Isocyanine-cured sealants, which are made of such oligomers, are oil-and-gasoline resistant in the range of requirements to Thiokol compositions.

It is also possible to add monomers and oligomers with other reactive groups together with dihalogenides during synthesis [2, 6, 30–32]. Such a chemical modification of liquid thiokols makes it possible to obtain sealing materials with better properties than those of conventional commercial thiokols.

Publications [33, 35] demonstrate a possibility of producing acrylonitrile- and styrene modified thiokols in conditions of active disulfide-disulfide exchange and assuming their addition during synthesis [33–35]. The analysis of infrared spectra of styrene-modified liquid thiokols has shown, that about 50% of added styrene is chemically bonded with a copolymer. Forming oligomers have low number-average functionality, which depends on the content of introduced styrene and the ratio of organic (styrene and formal) and inorganic ($Na_2S_x$) reactants. Synthesized oligomers are less vulcanizable, than industrial ones due to their lower functionality and are not suitable for exchange reactions. However, thiokol sealants, if modified by such oligomers, obtain higher adhesion to construction materials.

The synthesis of copolymer thiokols with unsaturated compounds as co-monomers is not so widespread probably due to a low selectivity of the process and corresponding weak physico-mechanical properties of related sealants.

The conventional method of Thiokol production leads, first of all, to formation of considerable amount of waste. However, the commercial production of Thiokol in source countries is carried out via reducing split of high-molecular polysulfide in dispersion (using sodium disulfide in Japan and Germany and sodium tetra sulfide in Russia). Table 1.4 shows existing facilities for production of liquid thiokols in various countries by 1990 and their range application [1].

**TABLE 1.4** The Economy of Liquid Thiokol: World Production by 1990 in Tonn

| Continent, Country | Production facilities | Total consumption of thiokols | Main application | | |
|---|---|---|---|---|---|
| | | | glass packets | Aircraft | Construction |
| America, USA | 27,000 (68%) | 5000 (15,9%) | 3000 (20,7%) | – | 2000 (46,5%) |
| Europe, Eastern Europe | 4000 (10%) | 15,000 (47,6%) | 10,000 (72,5%) | 4200 (33,0%) | 300 (7,0%) |
| USSR (former) | 4000 (10%) | 6500 (20,:%) | 500 (3,4%) | 4000 (31,5%) | 2000 (46,5%) |
| Asia, Japan | 4700 (12%) | 4000 (12,7%) | 500 tons 3,4% | 3500 (27,5%) | – |
| Other countries | – | 1000 (3,2%) | – | 1000 (8,0%) | – |
| Total | 39,700 (100%) | 31,500 (100%) | 14000 (100%) | 12,7000 (100%) | 4300 (100%) |

The analysis of world application of thiokols demonstrates that almost 60% of thiokol produced in 1990 was consumed by construction industry, where it was used for glass packets. The part of produced thiokol used for these purposes increased up to more than 80% in 2002. It seems, that it was the large amount of waste produced by "Morton International Inc." (USA) that led to liquidation of liquid thiokol production facilities in 2001. The world production of liquid thiokols made up 24,000 tons in 2002 and was concentrated in two countries only Germany (Akcros)–12,000 tons, Japan (Toray) – 8000 tons and Russia (OJSC Kazanskij Zavod SK)–1440 tons. The rest 2560 tons were unsold reserves of "Morton International Inc." [36]. To our opinion, there is no adequate replacement for liquid thiokols today in such areas as construction (production of glass packets, especially gas filled ones) and aircraft. Therefore, waste less liquid thiokol production technologies will be in demand in the coming years.

Besides liquid thiokols being limitedly produced due to ecological causes, there are other oligomers with end SH-groups being produced today having all sorts of the main chain's nature. There is limited information about oligomer captions and methods of their synthesis [5, 6, 37–39], but these compounds are of interest as they can form compositions with a substantially different chemical structure comparing traditional PSO.

The term polymer caption initially stood for a group of polymers produced by "Diamond Alkali Company" [1]. A. Damusis and E.M. Smolin proposed the following determination for polymer captions: "Polymer captions are in some respects intermediates between polysulfides and polyurethanes. The difference between polymer capstans and polysulfides is that first ones do not have sulfur or disulfide bonds in an elementary macromolecular unit." Polymercaptans are polymers based on acrylate, butadiene, butadiene-acrylonitrile, propylene oxide, urethanes, chloroprene, etc., having aliphatic, cycloaliphatic or aromatic fragments and at least two mercaptan end groups.

The family of compounds with SH-end groups also includes thio polyesters and polyenethiol compounds, which can vulcanize like liquid Thiokol without heating.

Oligomer capstans are known to be synthesized by radical polymerization of unsaturated monomers in the presence of such telogens as thiols and xanthogendisulfides so as hydrogen sulfide [40, 41].

The most famous and widely applied representative of polyenepolythiols or polyesterpolythiols is PM polymer by Phillips Petr company (Belgium) [42]. PM polymer is polypropylene oxide with end SH-groups. It is synthesized by etherification of a-mercaptopropionic and dithiopropionic acids.

$$\text{HS - X (OCH}_2 - \overset{\overset{\displaystyle CH_3}{|}}{C})_{5-50} \text{-O-X-SH,}$$

Where

$$X = \quad \text{~CH}_2\text{CH}_2 \text{ -COONH}$$

There is an oligomer marked ZL-616, which is based on propylene oxide oligoether and synthesized by interaction of a polyoxipropylene-glycol prepolymer with isocyanate end groups by mercaptoalcohol [27]. Resulting oligomers contain 2.3 weight % of sulfur and 2.3 weight % of SH-groups.

ZL-616 based sealants possess good physic-mechanical properties, small residual deformation and good acid resistance but are not resistant to organic solvents. This property makes it possible to introduce lots of plasticizers, oils (castor oil, dioctylphthalate etc.), filling materials and produce inexpensive sealants for construction. For more information about how properties of PM-polymer ZL-616 based sealants depend on their composition see [2, 5].

VNIISK in S.-Petersburg (former Leningrad) has worked out the method of synthesis of TP-polymer of polythiopolyolpolyester, which has the following structure [43]:

$$CH_2-(O-CH_2-CH)_m-(O-CH_2CH_2)_p-O-CH_2-CH-CH_2-SH$$
$$|\qquad\quad CH_3 \qquad\qquad\qquad\qquad\qquad OH$$

$$CH-(O-CH_2-CH)_m-(O-CH_2CH_2)_p-O-CH_2-CH-CH_2-SH$$
$$|\qquad\quad CH_3 \qquad\qquad\qquad\qquad\qquad OH$$

$$CH_2-(O-CH_2-CH)_m-(O-CH_2CH_2)_p-O-CH_2-CH-CH_2-SH$$
$$\qquad\qquad CH_3 \qquad\qquad\qquad\qquad\qquad OH$$

where $m = 16 \div 27$, $p = 0 \div 3$.

Chlorine-containing polyoxypropyleneoxyethylenepolyol marked "Laprol" with MW of 1000,6000 and chlorine functionality of 2.1,6.0 was used as initial PE [44]. Synthesized oligothiol is a low-viscous liquid ($1.0 \div 3.0$ Pa·s at 20°C), which can be vulcanized by all liquid thiokol hardeners. Its vulcanizates have good set of properties, but there is one essential disadvantage, that is intensive swelling in water because of a hydroxyl group in b-position to sulfuric one, that excludes their application for outdoor sealants.

Another prospective oligomers are polysulfide polyesters with MW of $600 \div 20,000$ and the following monomer structure: $2,95\%$ $(R'O)_n$, where R¢ is alkylene $C_{2-4}$, $n = 6 \div 200$; $3,70\%$ $C_2H_4OCH_2OC_2H_4S_x$ and $1,50\%$ $[CH_2CH(OH)CH_2(S)_x]$, where $x = 1,5$, polysulfide contains thiol groups- $C_2H_4OCH_2C_2H_4SH$ and/or-$CH_2CHOFSH_2SH$ at both ends [45].

Such oligomers are synthesized by a method described in Ref. [43]. Produced polysulfide-polyester contains 7% of sulfur and 5.7% of SH-groups. Sealants based on such oligomercaptans possess good physico-mechanical properties, water resistance and satisfactory stability in solvents.

A new family of polysulfide rubbers has been recently synthesized. It is marked "Permapol" (P-2, P-3, P-5) [1, 46]. PRC (Products research and Chemicals Corporation (USA)) was the first company to have succeeded in production of thiokols. Permapol P-2 is an oligomer with end SH groups, the method for its synthesis is the same than for ZL-616 [47]:

$$HS-R' $$

$$>CHCH-O-CH_2CH-O-CH_2CH-R'-SH$$

$$HS-R' \quad CH_3 \quad CH_3 \quad CH_3$$

where

$$R' = -O-C-N-\overset{CH_3}{\diagup}$$
$$\qquad \overset{\|}{O}\overset{|}{H} \qquad N-C-O-CH_2CH_2CH_2-$$
$$\qquad\qquad\qquad \overset{|}{H}\overset{\|}{O}$$

The advantage of P-3 type polythioesters over traditional thiokol sealants consists in the absence of weak disulfide and formal structures. Such polyesters can have end groups of a various nature [1, 48]:

$$R-(-O-CH_2CH_2-S-CH_2CH_2-O-CH-CH_2-S-CH_2CH_2-O-CH_2CH_2-S-CH_2CH_2)-R$$
$$\qquad\qquad\qquad\qquad\qquad\quad CH_3$$

Permapol-P-2 based sealants are recommended for sealing of glass packets. P-3 based sealants possess increased resistance to ozone, UV radiation, fuel, they have 2 times better strength properties, 1.8 times less fuel vapor and inert gas permeability, better compatibility with traditional plasticizers, two times better adhesion to many substances, even after impact of fuel-containing aromatic hydrocarbons. They are used in aircraft for anticorrosive treatment of fuel tanks, etc.

Permapol P-5 is the mixture of ordinary "LP" series liquid thiokol with the product of ins interaction with dithiols, such as dimercaptodiethyl-sulfide in the ratio of 1:1–1:2 [49]. Sealants based on such oligomers are highly resistant to ozone, ultraviolet radiation and fuel, are two times more robust and possess less fuel vapor and inert gas permeability, than liquid thiokol sealants.

## 1.2 THE PROPERTIES OF LIQUID THIOKOLS

Liquid thiokols are viscous liquids with consistency of honey, having a color from straw to dark brown, their value of viscosity is from 1.26 to 1.31 $g/sm^3$ and they have faint mercaptan odor. The pH value of their aqueous extract is from 6 to 8. The flow point of commercial grades of liquid thiokols depends on their molecular weight and is in a range from +10°C for thiokols with a molecular weight of 8000 to –26°C for thiokols with a molecular weight of 1000 [50, 51].

LP (Toray Fine Chemicals Co., Ltd.) – Japan

| Parameter | LP-33 | LP-3 | LP-980 | LP-23 | LP-56 | LP-55 | LP-12 | LP-32 | LP-2 | LP-31 |
|---|---|---|---|---|---|---|---|---|---|---|
| Average molecular weight | 1000 | 1000 | 2500 | 2500 | 3000 | 4000 | 4000 | 4000 | 4000 | 7500 |
| The content of trichloropropane, % | 0.5 | 2.0 | 0.5 | 2.0 | 0 | 0 | 0.2 | 0.5 | 2.0 | 0.5 |
| Viscosity, Pa·s | 1.5–2.0 | 0.94–1.44 | 10–12.5 | 10–14 | 14–21 | 41–48 | 38–50 | 41–48 | 41–48 | 90–160 |
| The content of SH-groups, weight % | 5.0–6.5 | 5.9–7.7 | 2.5–3.5 | 2.5–3.5 | 2.0–2.5 | 1.5–2.0 | 1.5–2.0 | 1.5–2.0 | 1.5–2.0 | 0.8–1.5 |
| The content of moisture, weight % | <0.1 | <0.1 | 0.15–0.25 | 0.15–0.25 | 0.15–0.25 | 0.15–0.25 | 0.15–0.25 | 0.15–0.25 | 0.15–0.25 | 0.15–0.25 |

| OJSC "Kazanskij Zavod SK"–Russia. | | | | | |
|---|---|---|---|---|---|
| Parameter | NVB-2 | I | II | 32 | TSD |
| The content of trichloropropane, % | 2 | 2 | 2 | 0.5 | 0.5 |
| The content of SH-groups, weight % | 3–4 | 2.2–3.4 | 1.7–2.6 | 1.7–2.6 | - |
| Maximum content of moisture, weight % | 0.2 | 0.2 | 0.2 | 0.2 | 0.3 |
| Maximum content of total sulfur, weight % | 40 | 40 | 40 | 40 | - |

The properties of liquid thiokols produced in Germany, Japan and Russia are summarized in Table 1.5 [52–54].

**TABLE 1.5** Physico-Chemical Properties of Liquid Thiokols by Different Producers

| Thioplast G (Akzo Nobel Thioplast Chemicals GmbH & CoKG) – Germany | | | | | | | | | |
|---|---|---|---|---|---|---|---|---|---|
| Parameter | G-4 | G-44 | G-21 | G-22 | G-10 | G-12 | G-112 | G-1 | G-131 |
| Average molecular weight | 1000 | 1000 | 2500 | 2500 | 4000 | 4000 | 4000 | 4000 | 7500 |
| The content of trichlorpropane, % | 2.0 | 0.5 | 2.0 | 0.5 | 0 | 0.2 | 0.5 | 2.0 | 0.5 |
| Viscosity, Pa·s | 1.32 | 17.5 | 10–20 | 10–20 | 38–50 | 38–50 | 38–50 | 41–52 | 125 |
| The content of SH-groups, weight % | 6.8 | 5.75 | 2.6–3.1 | 2.6–3.1 | 1.6–1.8 | 1.6–1.8 | 1.6–1.8 | 1.7–1.9 | 1.15 |
| Density | 1.259 | 1.2 | 1.285 | 1.285 | 1.285 | 1.285 | 1.285 | 1.286 | 1.310 |

The combustion heat of thiokols is 24,075 kJ/kg, and the specific heat is 1.26 kJ/kg·K. The content of water in thiokol does not exceed 0.3 weight %.

The content of free sulfur in liquid thiokols does not exceed 0.1 weight %. Total sulfur content in Thioplast G and LP thiokols is 37–38%. The flash temperature exceeds 230°C, the glass-transition temperature is 60°C. Liquid thiokols are characterized by a high stability of properties, their warranty storage period is three years. Molecules of liquid thiokols are either linear or branched and, as a rule, trifunctional. The number-average functionality ($f_n$) of liquid thiokols depends both on molecular weight and the quantity of a branching agent. Its value is somewhat less than the theoretical one. Its value is 2.22–2.27 for NVB-2, 2.3–2.48 for the first grade thiokol and 2.33–2.68 for the second grade thiokol.

TCP does not seem to participate fully in polycondensation in industrial liquid thiokol production lines. It can be caused by various reactivity of chlorine atoms in TCP molecules, which in turn effects the degree of branching of oligomers [55].

The value of $f_n$ in some known cases was substantially lower, than the theoretical one [56]. This is the effect of macromolecules, which does not contain SH-groups [56] and, for example, can have chains with other functional groups as well as cyclic macromolecules always occurring in PSO samples. Table 1.6 Demonstrates data on the functional type distribution parameters (FTD) and the content of cyclic molecules in some commercial grades of polysulfide oligomers [57].

**TABLE 1.6** Functional Type Distribution in Liquid Thiokols

| Grade | $\bar{M}_n$ | $\bar{f}_n$ | $\bar{f}_w$ | The content of cyclic molecules | |
|---|---|---|---|---|---|
| | | | | % (mol.) | % (weight) |
| I | 2350 | 1.60 | 2.45 | 33 | 6,3 |
| | 2250 | 1.43 | 2.48 | 42 | 6,8 |
| | 3500 | 1.77 | 2.57 | 31 | 4,0 |
| II | 2520 | 1.51 | 2.58 | 41 | - |
| | 2300 | 1.68 | 2.31 | 28 | - |

Wide FTD of such oligomers is determined by substantial (up to 7–8 weight %) content of cyclic and highly branched molecules. The functionality of branched molecules can amount to 6–8. Interchain exchange reactions, occurring during synthesis of such oligomers, result in the most probable molecular weight distribution $M_w/M_n = 2$ [58].

Liquid thiokols belong to the group of statistically branched polymers with three-functional branching centers. Their MWD does not depend on a molecular weight and the degree of branching of oligomer. It is in the range of 2.4–2.6, despite easily occurring thiol-disulfide and disulfide-disulfide interchain exchange processes.

The variation of viscosity of some the most widespread liquid thiokol grades with temperature arepresented in Fig 1.2.

The mechanism of interchain exchange in liquid thiokols is ionic heterolytic splitting of disulfide bond. Its rate depends on the degree of polymer's "polysulfidity." Non-catalyzed exchange activation energy is 52.8 kJ per mole. The specific heat capacity of formal-based liquid thiokols is 1.53–1.59 J/(g*C) at 10°C. The value of solubility parameter for such thiokols is 18–18.4 $(MJ/m^3)^{0.5}$ [59, 60].

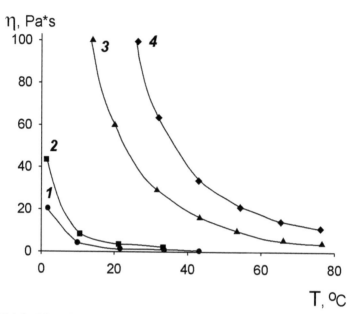

**FIGURE 1.2**  Viscosity-temperature dependence for liquid thiokols.

2,2¢-dichlorodiethylformal-based liquid thiokols and related vulcanizates (at stress-less state) do not crystallize during long-term storage at a low temperature.

Liquid thiokols are soluble in dioxane, benzene, chlorobenzene, toluene, dichloroethane, ethylenechlorohydrine, dibutylphthalate, phenol, resorcine, cyclohexanol, furfural, benzaldehyde and trichloroethylene. They demonstrate limited solubility in methylethylketone, xylene, diethyl ether, carbon tetrachloride, ethyl acetate and are insoluble in alcohols, ethers, glycols, kerosene, gasoline and water. Information on compatibility of liquid thiokols with various solvents is given in Table 1.7 [53].

**TABLE 1.7**  Compatibility of Liquid Thiokols with Various Solvents (25°C)

| Solvent | The quantity of solvent compatible with 100 parts of liquid thiokol | | | | | |
|---|---|---|---|---|---|---|
| | LP-32 | LP-55 | LP-12 | LP-2 | LP-31 | LP-3 |
| **Ketones** | | | | | | |
| Acetone | 70 | — | — | 80–125 | — | 400 |
| Methylethylketone | 200 | — | — | 300–350 | — | ∞ |

**TABLE 1.7** *(Continued)*

| Cyclohexanone | $\infty$ | $\infty$ | $\infty$ | $\infty$ | $\infty$ | $\infty$ |
|---|---|---|---|---|---|---|
| **Alcohols** | | | | | | |
| Methanol | 0 | 0 | 0 | 0 | 0 | 0 |
| Ethanol | 0 | 0 | 0 | 0 | 0 | 0 |
| Ethylene glycol | 0 | 0 | 0 | 0 | 0 | 0 |
| Furfuryl alcohol | — | — | — | 900 | — | — |
| Cyclohexanol | — | — | — | 25 | — | — |
| **Aldehydes** | | | | | | |
| Benzaldehyde | $\infty$ | $\infty$ | $\infty$ | $\infty$ | $\infty$ | $\infty$ |
| Furfural | $\infty$ | $\infty$ | $\infty$ | $\infty$ | $\infty$ | $\infty$ |
| **Ethers** | | | | | | |
| Diethyl ether | — | — | — | 30 | — | 70 |
| Dioxane | $\infty$ | $\infty$ | $\infty$ | $\infty$ | $\infty$ | $\infty$ |
| Methyl acetate | — | — | — | 80 | — | — |
| Ethyl acetate | — | — | — | 175 | — | — |
| Butyl acetate | 130 | — | — | 125 | — | — |
| Diethylphthalate | — | — | — | $\infty$ | — | — |
| Dibutylphthalate | — | — | — | $\infty$ | — | — |
| **Organic acids** | | | | | | |
| Acetic acid | 0 | 0 | 0 | 0 | 0 | 0 |
| **Aliphatic compounds** | | | | | | |
| Diluent naphta | 0 | 0 | 0 | 0 | 0 | 0 |
| Mineral spirits | 0 | 0 | 0 | 0 | 0 | 0 |
| **Aromatic compounds** | | | | | | |
| Benzene | $\infty$ | $\infty$ | $\infty$ | $\infty$ | $\infty$ | $\infty$ |
| Toluene | $\infty$ | $\infty$ | $\infty$ | $\infty$ | $\infty$ | $\infty$ |
| Xylene | 250 | — | — | 300 | — | $\infty$ |
| Phenol | $\infty$ | $\infty$ | $\infty$ | $\infty$ | $\infty$ | $\infty$ |
| Resorcine | $\infty$ | — | — | $\infty$ | — | — |
| **Chlorinated hydrocarbons** | | | | | | |
| Carbon tetrachloride | — | — | — | 200 | — | $\infty$ |
| Trichloroethylene | $\infty$ | $\infty$ | $\infty$ | $\infty$ | — | — |
| Perchloroethylene | 80 | — | — | 9 | — | — |

**TABLE 1.7**    *(Continued)*

| Nitroparaffins | | | | | | |
|---|---|---|---|---|---|---|
| 2-nitropropene | 600 | — | — | 470 | — | ∞ |
| Nitromethane | — | — | — | 150 | — | — |
| Nitroethane | — | — | — | 900 | — | — |
| 1-nitropropane | — | — | — | 900 | — | — |
| **1** | **2** | **3** | **4** | **5** | **6** | **7** |
| Acrylonitrile | — | — | — | 100 | — | — |
| Acetamide | 0 | — | — | 0 | — | — |
| Water | 0 | 0 | 0 | 0 | 0 | 0 |

$$HS-[R-S-S]_n-R-SH$$

$$HS-[R-S-S]_n-CH_2-\underset{\underset{S-S-[R-S-S]_m-R-SH}{|}}{CH}-CH_2-S-S-R-SH \qquad (A)$$

where $R = -CH_2-CH_2-O-CH_2-O-CH_2-CH_2-$

## 1.3    THE STRUCTURE OF LIQUID THIOKOLS

When TCP is used as a branching agent, ordinary branching in PSO can be accompanied by other mechanisms of branched structures formation [12, 13].

Realization of any of such mechanisms is stipulated by the interchain exchange rate, the content of impurities or the presence of free or total sulfur.

All the TCPs are considered to participate in formation of macromolecules with a statistical distribution of its links along a chain [9, 61].

PSO's fine structure has been studied by NMR $^{13}C$ method. Mazurek and Moritz have used this method to determine the quantity of TCP in liquid PSO [61]. The authors have concluded that the level of bound three-functional agent in some polymers is below its theoretical value, because TCP is already partly hydrolyzed in a reactor. Model compounds have been used to find deviations from the ideal chemical structure of PSO. In addition to the main resonance peaks (C–S, –C–O, O–C–O), other peaks have been found, which indicate the presence of C–OH-groups, small quantities of $-CH_2-OCH_2-OCH_2-OCH_2-$groups and their high molecular weight analogs.

Figure 1.3 shows NMR $^{13}$C spectra of linear and branched PSOs with various content of TCP. The structure of linear and branched oligomers, made of 2,2¢-dichloro diethylformal by the commercial production method, may look like this [2, 61]

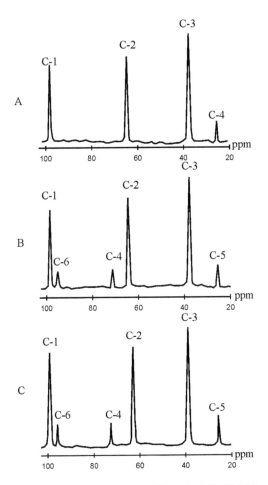

**FIGURE 1.3** NMR 13C spectra of linear (A) and branched (B, C) PSOs with TCP content of 3 (B) and 5 (C) mol. %.

Below are given chemical shifts and corresponding structural fragments of linear and branched PSOs.

| Carbon atom | C-1 | C-2 | C-3 |
|---|---|---|---|
| Functional group | $OCH_2O$ | $OCH_2$ | $CH_2S$ |
| $\delta^{13}C$ | 96 | 65 | 39 |

Resonance signals of carbon nuclei with chemical shifts of 39 (C-3), 65(C-2) and 96 ppm (C-1) indicates groups $CH_2S$, $OCH_2$ and $OCH_2O$ correspondingly independently on TCP content.

When more TCP is added, new signals appear in NMR spectra, indicating carbon nuclei C-5, C-4 and C-6 with corresponding shifts of 23, 69–71 and 90–91 ppm. Their intensity is in symbasis to the quantity of a branching agent.

According to information in scientific literature [12, 63], branching's can in most cases shift signals to the lower field area. For example, substitution of alkyl radicals in saturated hydrocarbons causes 9–10 ppm signal shift to weaker fields. Following characteristics have been obtained for model compounds with similar structure: 66,7 ppm signal is indicative for CH-group of chloroisopropyle $CICH(CH_3)_2$ [64]. 40 and 46 ppm signals indicate $CH_2$ groups of ethyl chloride and propyl chloride correspondingly. These data let us make the following interpretation of signals obtained during the study of the initial model compound: 1,2,3-trichlotorpopane (Fig 1.4): $CICH_2$= 47 ppm = 59 ppm. One must note, that chemical shifts of OCH groups were as a rule within 55–66 ppm for industrial batches of TCP.

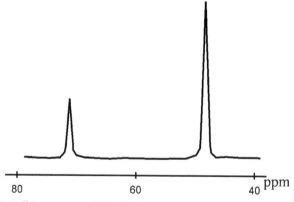

**FIGURE 1.4**   MNR $^{13}C$ spectrum of initial 1,2,3-trichloropropane.

The signal from a carbon nucleus C-5 with the shift of 23 ppm may correspond to the $-CH_2$ SH group, because the chemical shift of the end link in mercaptan $(CH_3)(CH_2)_{11}$ $CH_2$ SH has the value of 24,6 ppm [64]. When the content of TCP increases, the quantity of methylene groups, having shift of 23 ppm, may increase too. It is true for oligomers with a similar molecular weight.

The intensity of C-5 signal in spectra really grows in proportion to the quantity of TCP, introduced into PSO oligomer chain. Signals with chemical shifts of 71 and 91 ppm can be observed, as a rule, in branched oligomers when the quantity of a branching agent in increased (beyond 2 mol. %).

Appearance of carbon nuclei signals C-4 (71 ppm) and C-6 (91 ppm) can be stipulated by formation of long-chain branching's in oligomer chain, the influence of cycle formation processes and the presence of labile sulfur atoms, being able to absorb sulfur-containing groups of oligomer fragments. It is usually assumed that branched oligomers are present in PSO and have the above given structure (A). However, detailed study of PSOs reveals a broader set of possible structures of branched oligomers in liquid thiokols.

The statistical analysis of NMR $^{13}$C and $^{1}$H spectra of industrial and laboratorial PSO shows, that the pattern of observed spectra, the intensity of hydrocarbon and proton spectral signals, the presence and scattering of chemical shifts of specific oligomer fragments are determined by several reasons. These are, first of all, the storage time, the content of impurities as well as of total and unbound sulfur in PSO. The stated reasons are known to exert a strong influence on the occurrence and rate of thiol-disulfide and disulfide-disulfide interchain exchange in PSOs [2, 11]. The presence of labile sulfur and exchange reactions in the oligomer volume are also known to change the activity of oxygen-containing formal groups, $-OCH_2O-$ and $-CH-$ groups of propane link, which is introduced into oligomer chain [13, 62, 65]. Impurities and electron-donating sulfur atoms increase activity of $-CH-$ protons that could result in formation of new structural and branched fragments in an oligomer chain. Figure 1.5 shows the typical example of NMR $^{13}$C spectra. NMR $^{13}$C spectra, recorded at insufficient suppression of spin-spin interaction with protons show signals with the chemical shift of 71 and 91 ppm in a form of resonance triplets, indicating that these carbon atoms belong to $CH_2$-groups.

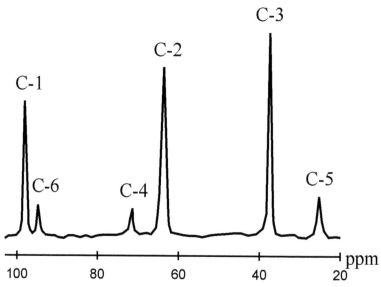

**FIGURE 1.5**   MNR $^{13}$C spectrum of industrial PSO (the content of TCP is 2 mol.%).

The occurrence of exchange reaction and increased activity of protons in propane fragments of oligomer chains could lead to formation of new structures or to reorganization of old branched fragments. One of possible alternatives is the formation of branching with the following structures:

$$-S-CH_2-\underset{\underset{S-S-R-SH}{|}}{CH}-CH_2- \qquad\qquad \begin{array}{c} S-S-R-SH \\ | \\ -S-CH_2-C-CH_2-R-SH \\ | \\ S-S-R-SH \end{array} \qquad (1)$$

$$\begin{array}{c} S-S-R-SH \\ | \\ -S-CH_2-C-CH_2-R-SH \\ | \\ SH \end{array} \quad (2) \qquad\qquad \begin{array}{c} SH \\ | \\ -S-CH_2-C-CH_2-R-SH \\ | \\ SH \end{array} \quad (3)$$

Considering the activity of SH-groups in conditions of PSO production, a partial cyclization of forming branched fragments (1) is also possible. It looks like this:

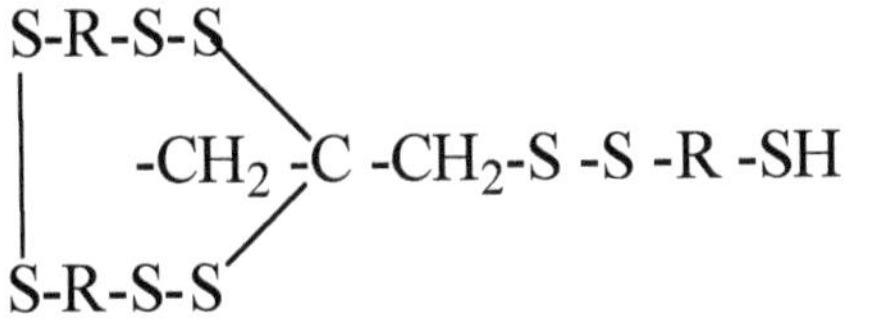

$$
\begin{array}{l}
\text{S-R-S-S} \\
\quad\quad\quad\diagdown \\
| \quad\quad -CH_2-C-CH_2-S-S-R-SH \\
\quad\quad\quad\diagup \\
\text{S-R-S-S}
\end{array}
\qquad (4)
$$

or formation of hydrogen bonds

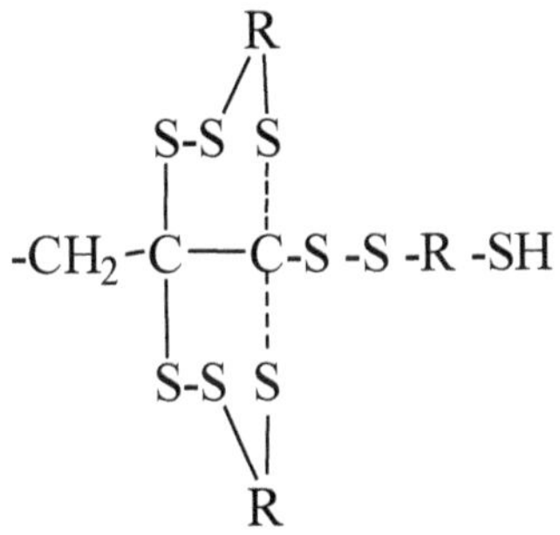

$$
\begin{array}{l}
\quad\quad\quad R \\
\quad\quad\quad /| \\
\quad\quad S\text{-}S\ \ S \\
\quad\quad | \quad\ | \\
-CH_2\text{-}C—C\text{-}S\text{-}S\text{-}R\text{-}SH \\
\quad\quad | \quad\ | \\
\quad\quad S\text{-}S\ \ S \\
\quad\quad\quad\ \backslash| \\
\quad\quad\quad\ R
\end{array}
\qquad (5)
$$

Cyclization of branching and formation of stable hydrogen bonds with –CH groups of propane fragments or with OCHO-groups of formal chains of oligomers is carried out by either intramolecular interaction of branching's or interaction of branching's, belonging to neighboring PSO macromolecules. Therefore, really observed C-4 signals with the chemical shift of 70–71 ppm can indicate the following structure:

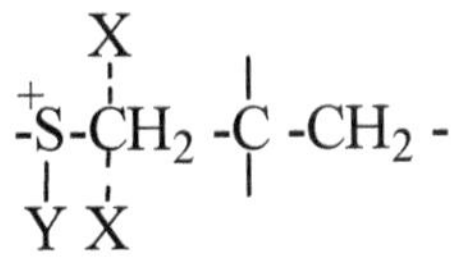

$$
\begin{array}{l}
\quad\ X \\
\quad\ | \\
^{+}\text{S-}CH_2\text{-}C\text{-}CH_2\text{-} \\
\quad | \ | \\
\quad Y\ X
\end{array}
\qquad (C\text{-}4)
$$

X = -S;  -S-S;  -SH;  Y-S.

Quaternary carbon signals are not present in spectra due to a low sensitivity, because of small quantities of such a fragment and absence of overhauser effect for non-protonated carbon atoms. Formation of hydrogen bonds is in all cases stipulated by the presence of liable sulfur atoms giving a positive charge to a sulfur atom in a linear oligomer chain and increasing mobility of $CH_2$-group atoms in a residual propane fragment. A relatively high mobility of hydrogen atoms results in coupling of a stationary electron pair in sulfur atoms of $SCH_2$–fragments with the highest valent orbitals of labile sulfur atoms, adsorbed along oligomer chain. It is confirmed by a paramagnetic shift of $(SCH_2)$-group's resonance signal in NMR spectra, whose chemical shift can usually be within 0.5–0.6 ppm depending on the content of total sulfur and added TCP.

Forming cyclic structures are unstable and can open with the course of time due to influence of various factors and contribute to the density of network chains during vulcanization of PSO.

Therefore, correlation of C–4 and C–6 signals in NMR spectra demonstrate difference from the usual picture for a branched oligomer. It was Shlyahter [61, 66, 67], who noted earlier about possible influence of TCP on the style of branching in liquid thiokols. All the mentioned set of structures seems to form in real conditions. Therefore, added TCP does not fully participates in formation of long-chain branching's during synthesis of liquid thiokols. The presence of a secondary chlorine atom in the TCP molecule, which is less reactive than primary atoms, leads to a certain decrease of the branching degree of oligomers, comparing theoretical values, and to emerging of fragments with new branched structures and HS-groups at methine carbon atom. As the quantity of TCP in the mixture of monomers increases, such groups become more important with chlorine atoms, not participating in PSO oxidation, and secondary HS-groups seem to be less active during vulcanization. Partial TCP hydrolysis and formation of new branched structures with reduced reactivity, as well as the increase of secondary HS-groups content and possible cyclization of PSO molecules during synthesis can cause substantial differences in experimental and calculated values of the chemical bond network densities, as it shown in Table 1.8 for unfilled vulcanizates.

**TABLE 1.8**   The Influence of TCP Content in Liquid Thiokols on the Degree of Conversion of HS Groups (W) and the Density of Vulcanization Network Chains

| Vulcanizing agent type | The quantity of TCP, % | The density of chemical chains in a network $v \cdot 10^4$ mol/sm³ | | W, % |
|---|---|---|---|---|
| | | experimental | calculated | |
| | 0.5 | 0.53 | 0.57 | 93 |
| | 1.0 | 0.56 | 1.15 | 48.6 |
| | 1.5 | 0.85 | 1.73 | 49.0 |
| $Na_2Cr_2O_7$ | 2.0 | 0.88 | 2.30 | 38.0 |
| | 3.0 | 1.04 | 3.46 | 30.0 |
| | 4.0 | 1.09 | 4.60 | 23.7 |
| | 5.0 | 0.90 | 5.80 | 15.58 |
| | 0.5 | 0.56 | 6.57 | 99.0 |
| | 1.0 | 0.81 | 1.15 | 71.0 |
| Paste $MnO_2$ 9 | 1.5 | 1.04 | 1.73 | 60.0 |
| | 2.0 | 1.14 | 2.30 | 49.0 |
| | 2.5 | 1.32 | 2.88 | 46.0 |
| Powder $MnO_2$ | 3.0 | 1.35 | 3.46 | 39.0 |
| | 5.0 | 1.24 | 5.80 | 21.4 |

## 1.4   INTERCHAIN EXCHANGE REACTIONS IN POLYSULFIDE OLIGOMERS

A characteristic feature of PSOs is their ability to enter into interchain exchange reaction between end HS-groups and di- or polysulfide main chain groups, as well as polysulfide groups with various degrees of polysulfidity. The practical value of these reactions is the opportunity of PSO synthesis with a narrow molecular weight distribution, independent on their average molecular weight. Moreover, it is possible to produce oligomer with required degree of polymerization by blending various batches. The chemical activity of PSO macromolecules is stipulated by two main factors: the mobility and electron donating ability of hydrogen atoms in mercaptan end groups as well as reduced –S–S– bond energy in di- or polysulfide fragments. The last one set the basis for exchange reactions both in oligomer bulk and in the area of its contact with other compounds.

A principal possibility of exchange reactions occurrence in PSO is confirmed by reduction of polymer's molecular weight in the presence of

$S^{2-}$, $HS^-$, $HO^-$, $CN^-$ ions due to exchange interaction of these ions with a disulfide group [68–71].

Numerous studies dedicated to the interchain exchange phenomena in polysulfides differ in estimations of not only the intensiveness of driving forces but of the ability of such reactions to occur in relatively mild conditions [66, 72–87].

Thiol-polysulfide exchange (TPE) and disulfide-disulfide exchange (DDE) reactions can be expressed in the following general equations:

$$RSSR + R'SH \longrightarrow RSSR' + RSH$$

$$RSSR + R'SSR' \longrightarrow 2RSSR'$$

Studies of mechanisms of these exchange reactions have shown, that they mainly have ionic nucleophylic substitution mechanism [88, 89], but the free radical mechanism takes place at increased temperatures and with no added solvents [70, 90, 91]. Thiol-disulfide and disulfide-disulfide exchange in neutral or alkaline solutions occurs via breakage of –S–S– bonds, attacked by mercaptide-ion [90, 91]:

$$RS^- + R'SSR' \longrightarrow R'S^- + RSSR'$$

The active particle in an acidic media is a sulfenic ion [86]:

$$RS^+ + R'SSR' \longrightarrow R'S^+ + RSSR'$$

The addition of thiols in acidic medium inhibits disulfide exchange. The role of thiol is the transformation of active sulfenic ions into protons, which regenerate sulfenic ions in the slow reaction:

$$RSSR + H^+ \xrightarrow{slow} {\sim}RS^+ + RSH$$

$$RS^+ + R'SSR' \xrightarrow{fast} R'S^+ + RSSR'$$

$$RS^+ + R'SH \longrightarrow RSSR' + H^+$$

Quite different behavior is observed for exchange reactions in some media [90]. For example, exchange interaction of aliphatic thiols and disulfides in the medium of aromatic hydrocarbons is very slow even at 200°C. The activation energy of this reaction has the value of 117.6 kJ/mol, and its rate is proportional to thiol's square root concentration. One has proposed a radical exchange mechanism in nonpolar media to explain these facts [90, 91]:

$$RSSR \longrightarrow 2RS^{\cdot}$$
$$RS^{\cdot} + R'SH \longrightarrow R'S^{\cdot} + RSH$$
$$R'S^{\cdot} + RS^{\cdot} \longrightarrow R'SSR$$

$$R'S^{\cdot} + RSSR \longrightarrow R'SSR + RS^{\cdot}$$

The initiation stage is rated by the split of polysulfide bonds with the degree of sulfidity above two, which are always present in PSOs [69]. Further stage of the exchange process is a set of substitution reactions by active groups of $RS^-$ or $RS^{\cdot}$ radicals by oligomer's $-S-S-$, $-S_n-$, SH– bonds, whose reaction rate depends on the ease of formation of these groups and their concentration. The easiest splitting of polysulfide bonds takes place in the presence of organic compounds [74].

The ease of a homolytic decay increases with the number of sulfur atoms in polysulfide, because the resonance stabilization of forming sulfur radical increases due to spin interaction with the labile electronic system of neighboring sulfur atoms [92].

The presence of free or bound sulfur in polysulfide oligomers is the major factor, which determines their reactivity. The increase of sulfur content in PSO above theoretical level can lead to the increase of free or labile sulfur concentration [93]. The theoretical content of sulfur in industrial 2,2'-dichlorodiethylformal0based oligomers (98 mol. %) and 1,2,3-TCP (2 mol. %) is 38.6%.

The increase of sulfur content in PSO mainly influences chemical shifts of proton-containing groups, which are the closest to the sulfur atom in an oligomer chain. The observed dependence is similar to the one obtained for symmetric organic polysulfides $R-S_n-R$ when the number of sulfur atoms in a polysulfide chain increases up to ten. These facts help to identify possible localization sites for a free sulfur. According to the obtained data free sulfur atoms are mainly localized on sulfur-containing fragments of PSO's main chain that are $SCH_2$ and HS-groups. HS-group of an oligomer has the strongest ability to localize free sulfur atoms. It is explained by a relatively high mobility of hydrogen atoms, causing conjugation of unpaired electrons of electronegative sulfur atoms in HS– and $SCH_2$ groups with highest valent orbitals of sulfur atoms in free molecules. The presence of free sulfur atoms, adsorbed along oligomer's chain, leads to a substantial decrease of sulfur atoms' electronic cloud density and the mobility of HS-protons, comparing oligomers with less content of free sulfur. It can mainly explain the increase of PSO's reactivity if few free sulfur atoms are introduced, that would increase the activity of mercaptan groups during curing of oligomers. Authors of [94, 95] observed a similar effect of proton activation in mercaptan groups in the presence of sulfur atoms.

The localization of free sulfur atoms by disulfide bonds or bonds with the degree of sulfidity above two, which are always present in PSOs, must accelerate exchange processes in oligomers and exert a significant influence on the modification of PSO or regulation of its structure as well as on making of copolymer oligomers. The effect of oligomer's reactive end groups activity increase with more free sulfur is necessary to consider during vulcanization of polysulfide oligomers and their compositions. The increase of content of sulfur atoms in PSO can also influence the stability of thiokols and their vulcanizates at service and long-term storage [96].

Exchange reactions, involving elementary sulfur, attract vast interest. Authors of [73, 96] have shown, that sulfur is easily soluble in aqueous solution of sulfides. Reactions can be summarized in the following equations:

$$HS^-_9 + S\begin{matrix} S-S-S \\ \diagup \quad\quad \diagdown \\ \quad\quad\quad S \\ \diagdown \quad\quad \diagup \\ S-S-S \end{matrix} \rightleftharpoons HS^-_{17}$$

$$HS^-_m + HS^-_n \longrightarrow HS^-_{m-j} + HS^-_{n+j}$$

Shlyahter with colleagues has studied molecular weight distribution (MWD) of polysulfides with mercaptan end groups and has specified, that thiol-disulfide exchange takes place not only during synthesis of PSO, but during mixing of oligomers with different molecular weight as well [87]. This is the fact, confirming thermodynamic nature of this phenomena.

Fraction analysis of oligomer mixtures shows, that mixing increases the number of fractions with average molecular weight. Differential MWD curves for individual oligomers and their mixtures indicate chemical interaction between oligomers, causing narrowing of MWD after mixing. It contradicts with experimental data in Ref. [87], where authors has shown, that MWD widens after mixing of oligomers. The last conclusion seems to have been made on uncertain results, because non-equilibrium oligomer mixtures have been used there.

Tobolskij [72, 73, 78] has demonstrated that thiol-disulfide exchange in bulk oligomer requires considerable activation, because disulfide bond is rather stable in these polymers. The values of exchange reaction rates, provided by different authors [73–75, 98, 99], are substantially different. The process of exchange interaction was sometimes easy, however, it required considerable activation in other cases [92]. Such a difference seems to be stipulated not only by the presence of impurities in thiokol, but by the structural features of studied PSOs as well. According to Ref. [74], the presence of impurities in PSO's is a rather important fact. Exchange processes in PSOs are substantially influenced by ions, such as alkali metals mercaptides and Lewis acids [77, 78], such as $FeCl_3$ with trace water. This initiating pair ($Fe_3^{++} H_2O$) is known to be more or less always present in industrial oligomers. Thiol-disulfide exchange accelerates under ultraviolet radiation [70].

The influence of structural differences is, first of all, reasoned by the degree of oligomer's polysulfidity. The presence of bonds, whose value of sulfidity exceeds two, facilitates exchange processes, because these bonds open much easily than disulfide ones [72, 73]. The rate ratio TDO DDO is within 1:2–1:8 here and depends on the structure of R and R' radicals, formed by sulfur atoms $S_n$ and ~R–SH bonds [90, 98, 99].

Activators of exchange processes, such as sulfur and bases, promote PSO vulcanization and Therefore, the problem, concerning correlation between vulcanization and exchange reactions is of vast importance [74].

Oxidation of thiol groups by polysulfide sulfur contributes to the process of structuring. Chains become less mobile during vulcanization. Therefore, some thiol groups avoid cross-linking due to their shielding by macromolecular chains from the activity sphere of an oxidant. Intensive process behavior helps to avoid such phenomena, as thiol groups will continuously join to oxidizing centers. According to [98, 99], interchain exchange in polysulfide, including vulcanizates, is the process to restore system's equilibrium. This mechanism consists in maintaining the balance between the chain length distribution of molecules and the reactivity of these molecules.

The relative easiness of both heterolytic and homolytic decay of di- and polysulfide chain fragments under external conditions makes it possible to observe interchain rearrangements of PSO macromolecules even at room temperatures [98]. The process rate will depend on the rate of formation of active particles and their current concentration. When the $n$ value of $S_n-$ bonds increases, they become more inclined to homolytic decay [90]. That's why the probability of the radical mechanism is greater for oligomers with denser polysulfide chains, other conditions being equal [98, 99].

Therefore, all what was said above lets us conclude, that sulfur-containing compounds are rather highly reactive and are inclined to both heterolytic and homolytic decay. The behavior and intensity of exchange processes in PSOs and vulcanizates are stipulated by the oligomer chain structure, the temperature and polarity of medium (solvents or chemically active impurities). It affords ground for considering them rather prospective for tailored modification of various polymer materials and explains many structural peculiarities of oligomers.

## KEYWORDS

- oligomers
- polymers
- polysulfide
- properties
- structure
- synthesis

## REFERENCES

1. Lucke, H. (1994). Aliphatic Polysulfide. Monograph of an elastomer. Publisher Huthig & Wepf Basel, Heidelberg, New York, 191 p.

2. Averko-Antonovich, L. A., Kirpichnikov, P. A., & Smyslova, R. A. (1983). Polysulfide oligomers and related sealants (in Russian). Leningrad: Himija–128 p.
3. Smyslova, R. A. (1974). Liquid Thiokol Sealants (in Russian). Moscow: CNIIT Jeneftehim, 83 p.
4. Smyslova, R. A., Shvec, V. M., & Sarishvili, I. G. (1991). Application of Curable Sealants in Construction. Review of VNIINTI and on the economy of commercial construction materials (in Russian). Series 6, 2, 30 p.
5. Smyslova, R. A. (1984). Liquid Thiokol Sealants (in Russian). Moscow: CNIIT Jeneftehim. P.67
6. Li, T. S. P. (1995). Kauchuk i rezina (in Russian). 2. P. 9–13.
7. Minkin, V. S., Deberdeev, R., Paljutin, Ja., Khakimullin, F. M., & Yu, N. (2004). Industrial Polysulfide Oligomers: Synthesis, Vulcanization, Modification (in Russian), Kazan, Novoe znanie, 175 p.
8. Minkin, V. S., Nistratov, A. V., Vaniev, M. A., Khakimullin, Yu. N., Deberdeev, R. Ja., & Novakov, I. A. (2006). Synthesis Structure and Properties of Polysulfide Oligomers. A review collected articles from the Volgograd State Technical University, Series "Chemistry and Technology of Element organic Monomers and Polymer Materials" (in Russian) Volgograd, 3(1), 9–20.
9. Shljahter, R. A., Novosjolok, F. B., & Apuhtina, N. P. (1971). Kauchuk i rezina (in Russian) 2, 36–37.
10. Apuhtina, N. P., Shljahter, R. A., & Novosjolok, F. B., (1957). Kauchuk i rezina (in Russian), 6, 7–11.
11. Bertozzi, E. R. (1968). Rubb Chem and Technol. 41(1), 114–160.
12. Shljahter, R. A., Novosjolok, F. B. (1976). Polysulfide Rubbers In the book: "Synthetic Rubber" (in Russian). Ed. by Garmonov, I. V., Leningrad: Himija, 552–571.
13. Suhanov, P. P., Averko-Antonovich, L. A., Minkin, V. S., & Kostochko, A. V., (1996). Zhurnal prikladnojhim ii (in Russian). 69(1) 124–126.
14. Suhanov, P. P., Khakimullin, Yu. N., Averko-Antonovich, L. A., & Minkin, V. S. (1994). Transactions of International Conference on Caoutchouc and Rubber (in Russian), Moscow, Vol. 3, 152–158.
15. Dojka, M., Blasiak, I., & Kucharski, M. (1985). Przem. Chem. 64(3), 129–131.
16. Dachselt, Thioplaste, Leipzig, (1971), 164.
17. Patent 3046516, FRG (Germany) C07 C149/05, (1981).
18. Patent 2073032, Russia, C08, G75/16, C08K 3/30.
19. Gobran, R. H.,& Berenbaum, M. B. (1969). In Polymer Chemistry of Synthetic Elastomers Ed. Kennedy, J. P., Tornqwist, E. G. M., N. Y. Lond. 805–842.
20. Patent 2099361 Russia, MKI S 08 G 75/14, 75/16.
21. Ioffe, D. S., Gubajdullin, L. Ju., Liakumovich, A. G., & Gubajdullin, I. L. (1997). Transactions of the Conference "Production and Consumption of Sealants and Other construction Compositions: Present State and Prospective (in Russian).Kazan. 91
22. Patent 2154056 Russia, MKI S 07 S 319/22, 321/14, S 08 G 75/16.
23. Ioffe, D. S., Gubajdullin, I. L., Shumilina, T. N., Liakumovich, A. G., & Kauchuk i rezina (2002) (in Russian),5, 4–5
24. Kohizsager, S. N. (1982). Pop. Plast, 27(2), 3–5.
25. Hobbs, S. I. (1992). Proc. ADC Div. Pol. Mat Sci. and Eng. Fall Meeting, 67, 415.
26. Patent 51632 Bolgaria, MKI5 S 08 G 75/14, S 08 L 81/04.

27. Larsen, R. J. (1971). of the IRY, V. 5(2), 62).
28. Patent USA 3422077, (1969). S.A., 14, 58596
29. Halikova, G. R., Pavel'eva, N. P., Samuilov, Ja. D., & Paljutin, F. M. (2004). Abstracts of International Conference on Caoutchouc & Rubber (in Russian), M., 241–242.
30. Averko-Antonovich, L. A., & Muhutdinova, T. Z. (1978). In the Book: Synthesis and physical chemistry of polymers. Kiev: Naukova Dumka.11 (in Russian) P. 20–24.
31. Kirpichnikov, P. A. (1979). Vysokomolek Soed, (in Russian) T. 21 A(11) S. 2457–2468.
32. Markov, M., Mladenov, I., Todorova, V., & Hristova, D. (1983). God. Higher Chemical Technology Institute Burgas, 18(2a), 181–190
33. Averko-Antonovich, L. A., Nigmatullina, F. G., Lekonceva, A. P., & Hakimullin, Ju. N. (1985). Kauchuk i rezina (in Russian) 11. P. 46–47.
34. Todorova, D., Markov, M., & Mladenov, I.v. (1981). God Higher Chemical Technology Institute Burgas .2 28–34
35. Mladenov, I. T., Markov, M. K., Todorova, D. D., & Todorov, S. N. (1984). Kauchuk i rezina (in Russian)12 S.8–11
36. Oral Presentation "Akzo Nobel" Raw materials with future" at International Conference «Glasstec», Dusseldorf, (2002)
37. Patent Application 60–18687 Japan, MKI S 08 G 75/14.
38. Patent Application 60–35368 Japan, MKI S 08 G 75/16, S 09 K 3/10.
39. Patent 8701378 RST, MKI S 08 G 75/14.
40. Fokina, T. A., Apuhtina, N. P., & Klebanskij, A. L. (1971). Vysokomol. Soed. (in Russian), Series A, 13(9), 1972–1979.
41. Garbon, J. M., Bross, G. K., J. Polym. Sci., Polym.Chem. Edit. (1976), 14(1), 159–181
42. Panek, W., Adh. Age. (1974) 17(1), 25–27.
43. Author's Certificate 704120 SSSR, MKI S 08 G 75/00.
44. Author's Certificate 826723 SSSR, MKI S 08 G 65/32.
45. Patent 5393861 USA, MKI5 S 08 K 5/15, C 08 K 5/20.
46. Singh, H. (1987), Rubber World, 196, 5. 32. R.34–36.
47. Patent 3923748 USA, MKI S 08 G 18/04.
48. Patent 4366307 USA, MKI C 08 G 75/00.
49. US-PS 4.623.711 PRC, L. Morris, H., Singh (corresponds to Permapol P-5). (21.08.1985 / 18.11.1986).
50. Smyslova, R. A., Kotljarova, S. V. (1976) Reference manual on rubber sealing materials (in Russian) Moscow: Himija. 72 p.
51. Spravochnik rezinwika (1971) (Handbook for the Rubber Industry–in Russian). Ed. by Zaharchenko, P. I., Moscow- Himija. 608 p.
52. Aczo Nobel's data sheet.
53. Torey's data sheet.
54. "KZSK" OJSC data sheets (in Russian), www.kzck.ru
55. Henricks, P. M., Hewitt, I. M., Pussell, G. A., Sandhu, M. A., & Grashof, H. R. (1981) Macromolecules.-V.I4. 6, 1770–1775.
56. Minkin, B. C., & Zykova, V. V. (1984) Zhurnal prikladnoj spektroskopii (in Russian). Vol. 46. 2, 318–321.
57. Jentelis, S. G., Evreinov, V. V., & Kuzaev, A. N. (1985). Reactive oligomers (in Russian).-Moscow: Himija, 304 P.

58. Nasonova, T. P., Shljahter, R. A., & Apuhtina, N. P. (1971). Vysokomolekul. Soedin (in Russian). Series B, 13(9), 635–637.

59. Muhutdinova, T. Z., Averko-Antonovich, L. A., Prokudina, K. N., & Kirpichnikov, P. A. (1973). KSTU Transactions (in Russian) Issue 50. 141–145.

60. Van-Krevelen, D. V. (1976). Properties and Chemical Structure of Polymers (in Russian). Moscow-Himija 416 p.

61. Shljahter, R. A., Nasonova, T. P., Apuhtina, N. P., & Sokolov, V. N., Vysokomol.soed (1972). (in Russian), Series. B, Vol.14, 1, 32–36

62. Mazurek, W., & Moritz, A. G. (1991) Macromolecules. V. 24. P. 3261–3265.

63. Levi, G., & Nelson, G. (1975). Manual on Carbon Nuclear Magnetic Resonance. (in Russian) 13, M., 295 p.

64. Formachek, V., & Desnoger, L. (1976). Data Bank 13C,–Copyring Bruker Physik., 1066 p.

65. Dojka, K., & Kuharski, M. (1980). Abstracts of 9th International Symposium on Sulfur Organic Compounds (in Russian), Riga, 81.

66. Nasonova, T. P., Kartasheva, G. G., Shljahter, R. A., Jerenburg, E. G., Vysokomol. Soed (in Russian) (1975). Series B, 17(2), 77–80.

67. Averko-Antonovich, L. A., & Kirpichnikov, P. A. (1978). Materials of the International Conference on Caoutchouc and Rubber (in Russian), Kiev, 44–51.

68. Fettes, E. M., & Iorczak, J. S., (1950). Ind. and Eng. Chem.22, 2217–2221.

69. Iorczak, J. S., & Fettes, E. M. (1951). Ind. and Eng. Chem., V. 43, 324–328.

70. Bertozzi, E. R., Davis, F. O., & Fettes, E. M. (1956). J.Polym. Sci., V.19, 9, 17–27.

71. Apuhtina, N. P. (1957). Chemical Science and Industry (in Russian), Leningrad, 323–330.

72. Owen, F. D., Macknight, M. I., & Tobolsky, A. V. (1964) J. Amer. Chem. Soc., V.68,4, 784–786.

73. Tobolsky, A. V., Macknight, M. I., & Takashi, M. J. (1964). Amer Chem. Soc., V.68, 4, ¤p.787–790.

74. Tavrin, A. E., & Guryleva, A. A. (1969) Izv. AN SSSR, the Chemical Series (in Russian), 6, 1300–1308.

75. Shljahter, R. A., Jerenburg, E. S., Nasonova, T. P., & Piskareva, E. P. (1965). Abstracts of International Conference on Macromolecular Chemistry, Praga, 412.

76. Fettes, E. M., & Mark, H. (1963). J. Appl. Polym. Sci. 7, 2239–2248.

77. Pryor, W. A. (1962). Mechanism of sulfur reactions, N.Y. Mc. Crow Ihill, 241p.

78. Mochulsky, M., & Tobolsky, A. F. (1948). Ind and Eng. Chem., 10, 2155–2163.

79. Golodny, P. L., & Tobolsky, A. F. (1959). J. Appl. Polym. Sci., 2, 39–45.

80. Minkin, V. S., Averko-Antonovich, L. A., Kachalkina, I. N., Kirpichnikov, P. A. (1975). Vysokomol. Soed. (in Russian), Series B,17(10),782–786.

81. Zykova, V. V. (1986). Chemical Sciences Candidate's Dissertation (in Russian), Kazan, KSTU.

82. Parker, A. I., Kharash, N. (1959). Chem. Rev, V.59, 583–629.

83. Berenbaum, M. V. (1967). Polysulfides Chemical reactions with polymers (in Russian), Moscow, 318–328.

84. Shaboldin, V. P., Demishev, V. N., Ionov, Ju. A., & Akatova, S. P. (1982). Vysokomol Soed. (in Russian), Series A24(5), 1099–1102.

85. Korotneva, L. A., Belonovskaja, G. P., & Dolgoplosk, B. A. (1968). Vysokomol. Soed. (in Russian), Series B, 10(1), 4–5.
86. Benesh, R. E., & Benesh, R. I. (1958). J. Am.Chem. Soc, 80, 1666–1669.
87. Shljahter, R. A., Apuhtina, N. P., & Nasonova, G. P. (1963). Dokl. AN SSSR (in Russian), Vol.149, 2, 345–347.
88. Parker, A. I., & Kharash, N. (1963). Chem.Rev, V.61, 644–649.
89. Foss, O., Kharash, N. (1961). Organic Sulfur Compounds, Organic Polymers, N.Y., 83–89.
90. Gur'janova, E. N., & Vasil'eva, V. N. (1954). Zhurnal fizicheskoj himii (in Russian), 28, 60–66.
91. Gur'janova, E. N., & Vasil'eva, V. N. (1955). Zhurnal fizicheskoj himii (in Russian), 29, 576–583.
92. Tavrin, A. E., Guryleva, A. A., & Dianov, P. P. M. (1967). Tejtel'baum B. Ja. DAN SSSR (in Russian), 174(1), 107–110.
93. Ivin, K. I., Lillie, E. D., & Peterson, I. H. (1973). Macromol. Chem. V.13, 217–240.
94. Corno, C., & Roggero, A. (1974). Eur. Polym J. 10(7), 525–528.
95. Nasonova, T. P., Shljahter, R. A., Novoselok, F. B., & Zevakin, I. E. (1973). Sintez i fiziko-himija polimerov (in Russian). Kiev, Issue 11, 60–63.
96. Nefed'ev, E. S., Khakimullin, Ju. N., Polikarpov, A. P., & Averko-Antonovich, L. A. (1986). Izv. VUZov (in Russian). 29(1), 97–100.
97. Shmidt, M. V. (1965). Interchange exchange Inorganic Polymers (in Russian), Moscow, 84–88.
98. Rozenberg, B. A., Irzhak, V. I., & Enikolopjan, E. S. (1975). Interchange Exchange in Polymers (in Russian), Moscow, 237 p.
99. Tobol'skij, A. V. (1964). Properties and Structure of Polymers (in Russian), Moscow, Himija, 93 p.

# CHAPTER 2

# VULCANIZATION, COMPOSITION AND PROPERTIES OF POLYSULFIDE OLIGOMER SEALANTS

## CONTENTS

## 2.1   INTRODUCTION

PSO-based composite materials (sealants), both single-component and two-component, usually contain the following components (below are listed the most widespread ones):

1. PSO – liquid thiokols or thiol-containing polyesters
2. plasticizers – dibutylphthalate, chloroparaffin, benzylbutylphthalate, etc.
3. fillers – chalk (natural occurring, chemically precipitated, or hydrophobic), kaolin, carbon black, titanium dioxide
4. thixotropic additives – aerosils, bentonits, hydrogenated castor oil
5. adhesive additives – epoxy resins, alkylphenolformaldehyde resins, functional trialkoxysilanes
6. promoters – sulfur, diphenylguanidine, thiuram, etc.
7. retarders – oleic, stearic or isostearic acids, metal stearates
8. drying agents (for single-component compositions) – ceolites, barium oxides
9. pigments – carbon black, titanium dioxide
10. vulcanizing agents – manganese and lead dioxides, sodium bichromate, sodium perborate monohydrate, calcium peroxide

There are lots of formulations for various applications, operating conditions coating technologies, etc. Basic principles of making compositions of sealants and their correlation with properties will be described below in this chapter using information from literature sources and by examples of author's research in chapters below.

The properties of PSO based sealants are determined not only by the structure of PSOs, but the nature of a vulcanizing agent, fillers and modifiers as well, while oxidation process rate is also stipulated by the presence of promoters, retarders and curing conditions (air temperature and humidity), in addition to the above listed factors. An end mercapto group contains mobile hydrogen atom. S-H bond dissociation energy (at kilojoules per mol) in organic compounds is small and substantially below than the corresponding value for O-H bond in alcohols [1]:

HS-H 373HO-H 486

$CH_3$S-H 373$CH_3$O-H 420

$C_2H_5$S-H 365$C_2H_5$O-H 415

Therefore, mercapto group splits homiletically substantially easily, than a hydroxyl group in corresponding alcohols. Sulfur in mercaptans is substantially more nucleophylic, than oxygen in alcohols, it is stipulated by both better polarizability of sulfur and the formation of hydrogen bonds in oxygen-containing compounds [1].

PSO vulcanization can be carried out in two ways [1–3]:
1. Oxidation of end SH-groups.
2. Interaction with compounds containing functional groups entering into the migratory polymerization reaction with end SH-groups.

Curing agents are subdivided into the following groups by nature [2]:
- Inorganic (for example, metal oxides);
- Organic (reactive oligomers, nitro compounds)
- Gaseous (air, oxygen)

## 2.2 VULCANIZATION OF POLYSULFIDE OLIGOMERS BY OXIDATION OF MERCAPTAN GROUPS

Oxidative curing of PSO at a room temperature is carried out by an oxidizing agent, which can be introduced into PSO or a composition with the second component (two-component system). Another way is when the oxidizing agent is initially added to a PSO blend, but such a blend must be prevented from the contact with an atmosphere. Reactive mercaptan groups of PSO turn into disulfide bonds during oxidation in the well-known reaction [1, 2]:

$$\text{R-SH} + 1/2\text{O}_2 + \text{HS-R} \longrightarrow \text{R-S-S-R} + \text{H}_2\text{O}$$

There are many inorganic and organic oxidants offered as oxidizing agents for a liquid thiokol [1, 2, 4, 5–10], including: in organic peroxides or dioxides ($\text{H}_2\text{O}_2$, $\text{ZnO}_2$, $\text{BaO}_2$, $\text{PbO}_2$, $\text{MnO}_2$, $\text{SbO}_2$, etc.), chromates or nitrates ($\text{Na}_2\text{Cr}_2\text{O}_7$, $\text{K}_2\text{Cr}_2\text{O}_7$, $\text{NH}_4\text{NO}_3$), monohydrates of alkali metals perborates, organic oxidants (quinoneoximes, nitro compounds, peroxides, hydro peroxides). However, the industrial application of hydro peroxides is limited by manganese and lead dioxides, sodium bichromate and sodium perborate for two-component sealants so as calcium peroxide and sodium perborate for single-component ones [1, 4, 11–12].

Each agent, used for vulcanization, adds specific properties to a sealant. This effect is stipulated by both the degree of PSO's curing and the influence of reduction products on the range of properties of vulcanizates [13, 14].

The data given in Table 2.1 provide a basic representation of advantages and disadvantages of vulcanizing agents and their applicability [2].

Industrial grade vulcanizing agents have the following row of activity [2, 4, 16, 17]:

$$PbO_2 > Na_2Cr_2O_7 > MnO_2$$

Systems with manganese dioxide are more preferable as their parameters of network formation are optimal. However, vulcanizates with sodium bichromate demonstrate higher mechanical, adhesion and cohesion properties, than vulcanizates, produced with metal dioxides.

Manganese dioxide is the most widespread and versatile oxidizing agent for two-component systems, because related sealants are the most resistant to light and UV-radiation, demonstrate good elastic recovery as well as deformation and strength properties (Table 2.1). They are also cost-effective and ecofriendly [1, 2, 4].

End SH-groups of oligomer are oxidized by manganese dioxide, form disulfide bonds and release water in the following common reaction [1, 18]:

$$2{\sim}R\text{-}SH + MnO_2 \longrightarrow {\sim}R\text{-}S\text{-}S\text{-}R{\sim} + MnO + H_2O$$

The question of oxidation mechanism (either ionic or radical) is still open. For example, alkalis and organic bases are supposed to favor dissociation of mercaptan groups [19]:

$$\text{-}R\text{-}SH + OH^- \longrightarrow \text{-}RS^- + H_2O$$

$$R'_3N\text{---}H\text{-}S\text{-}R{\sim} \longrightarrow R'_3N^+H + \text{-}RS^- \quad \text{или} \quad R'_3\overset{\overset{\displaystyle H}{\mid}}{N}\text{--}\overset{+\delta}{S}\text{-}\overset{-\delta}{R}{\sim}$$

**TABLE 2.1**  The Dependence of Properties of Liquid Thiokol Sealants on a Vulcanizing Agent Type

| Parameters | $ZnO_2$ | Cumene hydroper-oxide | $CaO_2$ | $PbO_2$ | $MnO_2$ | $NaBO_2$* $H_2O_2$* $H_2O$ |
|---|---|---|---|---|---|---|
| Elongation at rupture (DIN 52455) | 100–300% | 100–300% | 50–250% | 200–400% | 300–600% | 500–900% |
| Recovery (DIN 52458) | 50–70% | 70–85% | 50–70% | 70–80% | 80–95% | 80–95% |
| Shore A (DIN 53505) | 25–50 | 20–40 | 10–25 | 10–30 | 15–70 | 10–30 |
| Module at 100% elonga-tion and at 23°C (N/mm²) | 0.1–0.6 | 0.2–0.5 | 0.1–0.25 | 0.1–0.4 | 0.1–0.8 | 0.1–0.4 |
| Mainly for systems | 1-comp. | 2-comp. | 1-and-2-comp. | 2-comp. | 2-comp. | 1-and-2-comp. |
| Main applica-tion | glass cover | construction sealants | glass cover | con-struc-tion sealants | sealants for glass packets, chemically resistant sealants, rib-bons | construc-tion seal-ants, glass cover |

Exothermicity of the reaction provides high oxidation rate at a room temperature.

There are different viewpoints on interaction of metal oxides with PSOs. According to [20, 21], C-S-Me-S-C-type mercaptide bonds initially form:

$$2\text{-R-SH} + \text{MeO}_2 \longrightarrow \text{-R-S-S-R} + \text{MeO} + \text{H}_2\text{O}$$

$$2\text{-R-SH} + \text{MeO} \longrightarrow \text{-R-S-Me-S-R-} + \text{H}_2\text{O}$$

They are further oxidized by the extent of $MeO_2$

$$-R-S-Me-S-R- \ + \ MeO_2 \longrightarrow -R-S-S-R \ + \ 2MeO$$

or eliminated in thermal vulcanization or by sulfur:

$$-R-S-Me-S-R- \xrightarrow{t^0} -R-S-R \ + \ MeS$$

$$-R-S-Me-S-R- \xrightarrow{S} -R-S-S-R \ + \ MeS_0$$

Non-oxidized mercaptide bonds form less strong and thermally stable bonds in vulcanization network than disulfide ones and favor formation of cycles, the last effect is the most evident, when lead dioxide is used:

$$-S-S-CH_2 -CH_2 -O -CH_2 -O -CH_2 -CH_2 -S-S -CH_2 -CH_2 -O \rightleftharpoons$$
$$S-CH_2-CH_2-O-CH_2$$

$$-S-S-CH_2 -CH_2 -O -CH_2 -O -CH_2 -CH_2 -S^- \ + \ (S)-CH_2 -CH_2 -O$$
$$S-CH_2 -CH_2 -O \quad CH_2$$

It does not exclude that the process occurs involving mercaptide bonds of various chains:

$$\left.\begin{array}{l} \sim R\text{-}S\text{-}Pb\text{-}S\text{-}R\sim \\ \sim R'\text{-}S\text{-}S\text{-}R'\sim \end{array}\right\} \longrightarrow \begin{array}{l} \sim R\text{-}S \\ \quad | \\ \sim R'\text{-}S \end{array} + \left[\begin{array}{l} Pb\text{-}S\text{-}R\sim \quad или \quad PbS_2^+ \quad \sim R\text{-}(S)\text{-}R'\sim \\ S\text{-}R\sim \end{array}\right]$$

$$\left.\begin{array}{l} \sim R\text{-}S\text{-}Pb\text{-}S\text{-}R \\ \qquad\qquad\qquad S \\ \sim R'\text{-}S\text{-}S\text{-}R'\text{-}S \end{array}\right\} \longrightarrow \begin{array}{l} \sim R\text{-}S \\ \qquad Pb \\ \sim R'\text{-}S \end{array} + \begin{array}{l} S\text{-}R\text{-}S \\ | \quad | \\ S\text{-}R'\text{-}S \end{array} \quad или \quad \begin{array}{l} \sim R\text{-}S \\ |, | \\ \sim R'\text{-}S \end{array} + PbS_2 + \begin{array}{l} \sim R\text{-}S \\ | \\ \sim R'\text{-}S \end{array}$$

Authors of [23–26] have found that the process of vulcanization of PSO by metal oxides is heterogeneous and its rate is mainly determined by the degree of oxidizer's dispersity. Vulcanization rate is described as the sum of rates of heterogeneous reactions, taking place on the surface of crystalline oxidizing particles.

Taking into consideration adsorption phenomena, first reported by Tavrin [23], this reaction looks like this:

$$2\sim R\text{-}SH \ + \ MnO_2[MnO_2]_{cryst.} \longrightarrow \sim R\text{-}S\text{-}S\text{-}R\sim \ + \ MnO[MnO_2]_{cryst.} + H_2O \quad (1)$$

$$2\text{~R-SH} + \text{MnO[MnO}_2]_{cryst.} \longrightarrow \overset{\text{~RS}}{\underset{\text{~RS}}{|}}\text{MnO[MnO}_2]_{cryst.} + \text{H}_2\text{O} \quad (2)$$

Where [MnO$_2$] cryst. Is the manganese oxide crystallite, MnO$_2$ is the amorphized surface layer of [MnO$_2$] cryst. or defect areas inside a crystallite.

The rate of this process depends on adsorption-desorption equilibrium conditions [27–29]. Therefore, ion-exchange processes continuously occur in bulk PSO being cured and create electro activating zones for atoms with mobile electrons. A considerable part of atoms is immobilized by-S-Mn-S- bonds [30].

Elementary sulfur or thiokols macromolecules with polysulfide bonds, which are always present in industrial thiokols, lead to formation of sulfides on the surface of a crystalline metal oxide:

$$\overset{\text{~R-S}}{\underset{\text{~R-S}}{}}\text{Mn[MnO}_2]_{cryst.} + \text{S} \longrightarrow \text{~R-S-S-R~} + \text{MnS[MnO}_2]_{cryst.}$$

$$\text{~R-S}_x\text{-R~} + \overset{\text{-R-S}}{\underset{\text{-R-S}}{}}\text{Mn[MnO}_2]_{cryst.} \longrightarrow$$

$$\text{~R-S-S-R~} + \text{~R-S}_{x-1}\text{-R~} + \text{MnS[MnO}_2]_{cryst.}$$

Resulting structures contain disulfide bonds and provide high thermal stability of vulcanizate [1].

Latest studies concerning the mechanism of PSO's curing by manganese dioxide are given in Ref. [31]. EPR research has been done for the process of liquid thiokol oxidation by manganese dioxide at the presence of bases (such as diphenylguanidine), and it has been established, that this process generates thiyl radicals. The last ones have been proposed to interact with thiolate radicals and form anion radicals, which turn further into disulfides in the process of oxygen-mediated electron transfer.

Manganese dioxide is mainly used together with oxidation promoters tetramethylthiuram disulfide, diphenylguanidine, etc. The process of MnO2 oxidation by caustic soda is used extensively at the present time [32, 33].

Curing agents are usually applied in a form of vulcanizing pastes with such plasticizers as dibutylphthalate (DBF), benzyldibutylphthalate, chloroparaffin (CP), etc. to improve distribution of a vulcanizing group inside curing agent and increase conversion of end SH-groups during curing. [1, 34–36].

PSO is cured by manganese dioxide in the following reaction [11]:

$$2\text{~R-SH} + \text{PbO}_2 \longrightarrow \text{~R-S-Pb-S-R~} + \text{PbO}$$

Forming lead mercaptides are less thermally stable than, for example, zinc mercaptides and Therefore, mercaptide bond transforms into disulfide bond at the presence of sulfur easily enough:

$$\text{~R-S-Pb-S-R~} \xrightarrow{\text{t,S}} \text{~R-S-S-R~} + \text{PbS}$$

Lead dioxide is less toxic, and related sealants possess worse deformation and strength properties, than compositions with manganese dioxide. Therefore, it is not supposed to be used for large-scale production of PSO sealants despite its high activity to PSO.

Sodium bichromate is of vast interest as a vulcanizing agent. Professor Averco-Antonovich, L. A., has supervised a series of research works concerning the study of mechanism of PSO curing by sodium bichromate [9, 37–42]. Sodium bichromate enables to form networks with the highest density of chemical chains. Sodium bichromate oxidizes liquid thiokol in the reaction having the following mechanism:

$$2\text{Cr}^{6+} + 2\text{~R-S}^- \longrightarrow 2\text{Cr}^{5+} + \text{~R-S-S-R~}$$

$$\text{Cr}^{5+} + 2\text{~R-S}^- \longrightarrow \text{Cr}^{3+} + \text{~R-S-S-R~}$$

PSO is oxidized by of sodium bichromate in a water solution with occurring accompanying redox reactions [43].

The total electron transition $\text{Cr}^{6+}$ to $\text{Cr}^{3+}$ has been established to occur by the chronovoltam perometric method [40].

Sodium bichromate is used in production sealants in a form of aqueous or alcohol solution. Application of sodium bichromate results in promoting adhesion and strength characteristics of vulcanizates at ordinary temperatures. It is explained by the occurrence of donor-acceptor interactions along with covalent bonding of Cr atoms with S or O atoms on a polymer chain, whose role increases with the content of an oxidant [39]. Such bonds participate in formation of a network. It is proved by the reduction of effective network chains density after placing of vulcanizates made using sodium bichromate into water for two month (by 33 and 71% for unfilled vulcanizates and vulcanizates, filled with carbon black, correspondingly). The effective density of network chains has been established to decrease in sol-fraction free vulcanizate to 35–50% [44]. The sol-fraction in PSO vulcanizates is considered to mainly consist of low-molecular cycles (R)–S, wheren = 3–10 [22].

The MWD of PSO is close to the most probable value. Therefore, the considerable part of structural in homogeneities appears due to adsorptive and donor-acceptor interactions, taking place along with formation of chemical molecular bonds [45].

Zinc oxides and dioxides are used as vulcanizing agents [1, 2, 11, 46–49].

The mechanism of PSO curing by zinc oxide can be represented by the following equation:

$$2 \sim R\text{-}S\text{-}H + ZnO \longrightarrow \sim R\text{-}S\text{-}Zn\text{-}S\text{-}R\sim + H_2O$$

Addition of zinc oxides enables to obtain sealants of any color, containing minimal pigment quantity and having good physico-mechanical properties. Zinc dioxide ($ZnO_2$) is also good as a curing agent if active amines (such as ethylene diamine) are added [50]. Sealants with good physico-mechanical properties can be obtained by zinc oxide or zinc dioxide vulcanization of liquid Thiokol's in the presence of sulfur containing promoters. For example, combination of zinc oxide or dioxide with tetra methylthiuramdisulfide and sulfur results in formation of fast-setting Thiokol sealants with the strength of up to 5 MPa and the relative elongation of 1000%. The combination of zinc oxide with captax or altax makes it possible to produce fast-setting compositions for dental casts. ZnO vulcanization is effective at the presence of acidulous compounds [51–53] (levulinic acid, oligo chlorophos phazenes) and their combination with sulfur-containing organic compounds. It should be noted, that the color of zinc dioxide and zinc oxide based sealants is white, if pigments are not added (Table 2.1). However, their elastic recovery and relative elongation parameters are worse, than those of compositions, based on manganese dioxide.

Some attention is given to "melt"-type compositions, suitable for the high-temperature extrusion and making of extruded products for sealing glass and sashes. Zinc oxide (or dioxide) is the vulcanizing agent for such compositions, which are suitable for processing at 100–300°C with further curing and formation of coatings. Levulinic acid is proposed for a curing activator [52], so as various sulfur-containing promoters captax, ziram, thiuram, etc. [54–56]. Melts with such properties can be prepared via curing of liquid thiokol by tin di-n-butylate oxide [57].

Liquid thiokols are sometimes vulcanized by organic peroxides. This method can be used to produce sealants with white color. Isopropyl benzene hydro peroxide is the most widespread reactant for these purposes:

$$2\text{~R-S-H} + C_6H_5C(CH_3)_2\ COOH \longrightarrow R\text{-S-S-R-} + C_6H_5C(CH_3)_2\text{ -OH} + H_2O$$

Liquid thiokol with the molecular weight of 3000–4000 is supposed to be used here, as application of thiokols with molecular weight (1000), or taking peroxide in the excess, will make the vulcanization process too active [1]. As a result, oxidation will be is too deep with thiokol's main chain destructed to thiosulfonates or sulfonic acids by emerging free radicals [1]. Therefore, these vulcanizates are less thermally stable. Such sealants are blended with amines: 2,4,6-tris(dimethylaminomethyl)phenol, dimethylaminomethyl, tri-n-butylamine, diphenylguanidine, tetramethylguanidine [2] to prevent undesirable processes. Forming ion-radicals of amines bound radicals, which emerge as a result of organic peroxides decay, and prevent oligomer chains from destructing [1].

CA and Li-containing peroxides or their mixtures are used in industry along with Zn peroxides to make single-component thiokol curing blends together with other metal peroxides. Their major application is the production of single-component blended sealants [2]. Vulcanization of liquid Thiokol's by such blends is effective in humid air. Currently used compositions are usually represented by the following blends: $CaO_2/BaO$; $ZnO_2/BaO$; $ZnO_2/LiO_2$. Such curing agents carry out vulcanization in the following set of reactions:

$$CaO_2 \xrightarrow{H_2O} Ca(OH)_2 + O$$
$$2\text{-R-SH} + O \longrightarrow \text{-R-S-S-R-} + H_2O$$
$$2\text{-R-S-H} + ZnO_2 \longrightarrow R\text{-S-S-R-} + Zn(OH)_2$$
$$BaO + H_2O \longrightarrow Ba(OH)_2$$

Forming barium hydroxide effectively promotes vulcanization of liquid Thiokol's by zinc or calcium dioxides.

Another PSO oxidants, which have been recently discovered, are monohydrates of alkaline metal perborates (Na, K) [2, 58–60]: $NaBO_2$ * $H_2O_2$ * $H_2O$

They make it possible to produce construction sealants being highly resistant to air, UV-radiation and water, possessing excellent relative elongation and elastic recovery characteristics (Table 2.1). One can observe that application of monohydrates of alkali metal perborates enjoys stable growth in Western Europe today.

## 2.3  CURING OF POLYSULFIDE OLIGOMERS BY REACTIVE COMPOUNDS

The presence of SH-groups in PSOs stipulates the possibility of its entering into the migratory polymerization reaction with other monomers and oligomers, having reactive end groups. The network in such blends forms via catalytically activated interaction of oligomeric end groups, assuming at least one of them is branched. Thus, two processes occur simultaneously: PSO chemical modification and the curing of forming oligomer. Network density (the degree of curing) depends on the ratio of groups, which are able to interact in the presence of a catalyst. The complete curing state corresponds to absence of unreacted functional groups in a blend [61].

Such a curing method enables to create a family of sealing materials with reduced content of thiokol, whose characteristics overcome thiokol sealants at batch production. The best studied reactive oligomers in reactions with PSO are epoxy resins, isocyanates and acrylates [1, 4, 11, 61–64]. When oligomeric compounds are used, the main chain of PSO is enriched by blocks with comparable molecular weight. Sealants, made by copolymerization of PSO with unsaturated monomers, are interesting in this respect. There are novel researches on interaction of liquid thiokol with methylmetacrylate, tetraethyleneglycole dimetacrylate, etc. in the presence of free radical initiators, such as peroxides or hydroperoxides and activators such as amines [65–67]. Interaction of liquid thiokols with acrylates can be described by the following common reaction:

$$2\text{-R-SH} + H_2C=C-R'-C=CH_2 \longrightarrow \text{R-S-CH}_2\text{-CH-R'-CH-CH}_2\text{-S-R}$$

(with the pendant $CH_3$, $CH_3$ groups on the left reactant's central carbons labeled $CH_3$ $CH_3$, and on the product's central carbons labeled $CH_3$ $CH_3$)

The curing is radical and can be also accompanied by interchain exchange processes [68–71]. However, unsaturated monomers cause shrinkage of sealants and thus it is more preferable to use unsaturated oligomers, such as oligoesteracrylates (OEA) [72–74]. PSOs with acrylate end groups (ZL-2244 or ZL-1866) have been worked out quite recently. Their interaction with ordinary liquid thiokols results in formation of a new family of sealants [3].

Epoxy resins (ER) are the most widespread copolymerization agents for PSOs [75–81]. If the quantity of introduced ER is comparable to the quantity of PSO (30 mass parts and more), two parallel processes can take place [61, 63, 82]:

Formation of a block copolymer via interaction of end groups of polysulfide and epoxy oligomers:

$$H_2C-CH \sim HC-CH_2 + 2\sim\text{R-SH} \longrightarrow \sim\text{RS-CH}_2\text{-CH}\sim\text{CH-CH}_2\text{-SR}\sim$$

(with epoxide oxygens O bridging the left carbons, and OH OH on the product's central carbons)

Formation of interpenetrating networks as a result of independent epoxy resins and liquid thiokol's curing processes.

When liquid thiokol and ER undergo catalytic interaction at a room temperature, copolymers are formed under the influence of amino compounds. Their properties combine those of liquid thiokols, such as elasticity and resistance to aggressive media and ER's properties such as high adhesion to various substrates [1, 83]. The properties of such copolymers can vary in a wide range, depending on the ratio of used components.

The combination of equimolar quantities of sodium bichromate with monoethanolamine have been proposed for effective copolymerization of PSO with epoxy resin [84]. The curing mechanism of such composition has been studied. It has been determined, that ionic polymerization can be accompanied by formation of consecutive interpenetrating network of polysulfide and epoxy components, while the last one is much more defective [85]. The proposed curing systems are extremely active and make it possible to obtain fast-curing sealants with good range of properties.

There is the method of curing thiokol-epoxy composition at the room temperature and at the presence of tertiary amine catalysts [63, 86–88].

Liquid thiokol-ER copolymerization is highly effective when Mannich bases (MB) are used [63, 89]. It has been determined by EPR method, that curing of thiokol-epoxy compositions has the radical mechanism in the presence of MB [88–91]. MB homiletically decays and forms relatively stable nitroxyl radicals [92, 93], which are able to open epoxy ring and form alkoxyl radicals [88]:

$$R^{\bullet}_{in} + R' \text{-HC-CH}_2 \longrightarrow R' \text{-}\overset{\bullet}{C}\text{-CH}_2 \longrightarrow R' \text{-C-}\overset{\bullet}{C}H_2 + R' \text{-HC-CH}_2 \longrightarrow$$

$$R' \text{-C-CH}_2\text{-CH}_2\text{-CH-R'} + R''H \longrightarrow R' \text{-C-CH}_2\text{-CH}_2\text{-CH-R'} + R^{\bullet}''$$

where $R'_{in}$ are the radical products of Mannich base decay.

Forming radicals are consumed in homocuring of E-40 resin, as well as in the process of hydrogen atom removal from thiokol's SH groups, with further formation of copolymer structures [90].

$$R^{\bullet}_{in} + H_2C\text{-CH-R'-HC-CH}_2 \longrightarrow R_{in} H_2C\text{-CH-R'-HC-CH}_2 +$$

$$+ HSR''SH \longrightarrow R_{in} H_2C\text{-CH-R'-HC-CH}_2 + {}^{\bullet}SR''SH \longrightarrow$$

$$R_{in} H_2C\text{-CH-R'-HC-CH}_2\text{-SR''SH} + HSR''SH \longrightarrow$$

$$\longrightarrow R_{in} H_2C\text{-CH-R'-HC-CH}_2\text{-SR''SH} (1) + {}^{\bullet}SR''SH$$

$$+ H_2C\text{-CH-R'-HC-CH}_2 \longrightarrow$$

$$\longrightarrow (1) + HSR''\text{-S-} H_2C\text{-CH-R'-HC-CH}_2 \longrightarrow$$

$$R_{in} H_2C\text{-CH-R'-HC-CH}_2\text{-SR''S}^{\bullet} + HSR''\text{-S-} H_2C\text{-CH-R'-HC-CH}_2 \longrightarrow$$

$$R_{in} H_2C\text{-CH-R'-HC-CH}_2\text{-SR''S} H_2C\text{-CH-R'-HC-CH}_2\text{-S-R''SH} \text{ and etc.}$$

where $R_{in}^*$ are radicals, generated by a curing system

This fact is confirmed by the results of research in the model dodec-anethiol-phenylglycidyl ester reaction at the presence of Mannich bases as well as by $^{13}$C NMR spectra of thiol-epoxy copolymers [88].

Another method of sealant production has been proposed quite recent-ly. At first, adduct of PSO and epoxy resin is produced, then it is blended with more PSO and filler and subsequently cured. Resulting sealants have flaking strength of 3 kGs/sm [94]. Laproxides (oligooxypropyleneglycole-based epoxy resins) can be used for construction thiokol sealants [95].

Sealants obtained by the catalytic interaction of PSO with diisocya-nates are of both vast scientific and practical interest. When diisocyanate is introduced into PSO, thiourethane polymers form [96–100].

Copolymerization of PSO and isocyanates occurs in the presence of catalysts. If low-molecular weight diisocyanate is used for interaction with PSO, the synthesis is carried out at high temperatures or with catalyst, such as tin octoate or mercury phenylacetate [101]. Polyurethanes can be produced via joint curing of disulfide-containing polyol, diisocyanate and ethylenedimercaptan at 50°C and in the presence of antimony pentoxide [102].

Another method to produce thiourethane sealants is a non-catalyzed copolymerization of a prepolymer based on polyethers or polyesters, hav-ing isocyanate end groups with polyether-based oligomeric dimercaptan [103]. There is also the method of interaction of liquid thiokol with a prepolymer, which has isocyanate end groups, in the presence of tertiary amines, dibutyltindilaurinate [104].

Dimethylaminomethylphenols (Mannich bases (MBs)) and their trans-mination products, such as ethylenediaminomethyl substituted phenols (Agidol AF-2), turned out to be highly effective catalysts for PSO-diiso-cyanate copolymerization. Catalyst concentration, required for a notice-able catalytic effect, was 500 times less than the concentration of reacting end groups [99]. The degree of conversion of oligomers into thiourethane copolymer increases with the basicity of Mannich bases [105].

PSO-diisocyanate interaction can be considered to be a primary elon-gation of an oligomer chain with its further structuring when the content of isocyanate groups exceeds the content of mercaptan ones [61, 105, 106]:

$$\sim\text{R-SH} + \sim\text{R'NCO} \xrightarrow{\text{catalyst}} \sim\text{R-S-}\underset{\underset{O}{\|}}{C}\text{-NH-R'}\sim + \text{R'-NCO} \longrightarrow \sim\text{R-S-}\underset{\underset{\underset{\text{NH-R'}\sim}{|}}{\underset{C=O}{|}}}{\overset{\|}{C}}\text{-N-R'}\sim$$

The chain elongation reaction rate has been determined to exceed substantially the rate of secondary structuring reactions [61].

Authors of Ref. [87] have proposed thiourethanes to form via the catalyst-diisocyanate transition complex:

$$\text{RNCO} + (CH_3)_2\text{N-CH}_2\text{-Ph-OH(OM)} \rightleftharpoons \left[ \begin{array}{c} \text{R-N=C=O} \\ \uparrow \\ H_3C\text{-}\overset{|}{N}(CH_3)\text{-CH}_2\text{-Ph-OH} \end{array} \right] \longrightarrow$$

$$\xrightarrow{\text{R'-SH}} \text{R-N(H)-C(O)-S-R'} + (CH_3)N\text{-CH}_2\text{-Ph-OH}$$

In addition, formed thiourethane can contribute to the formation of transition complexes.

Toluylenediisocyanate (TDI) was used in Ref. [107] as a model system to study the structuring of PSO-macrodiisocyanate compositions. TDI is widely used to synthesize urethane prepolymers. PMR spectra of concentrated PSO-TDI solutions demonstrate a single within 10.6–10.8 ppm, with can be explained by formation of adducts.

$$\text{Ph-}\overset{+}{N}\text{ H=C=O}$$
$$\underset{\text{S-R}\sim}{|}$$

Temperature measurements have confirmed, that NCO- and SH-groups start to interact, forming labile bonds with lifetimes logically dropping with the decrease of medium's viscosity, as well as with the increase of temperature.

The curing mechanism for PSO-diisocyanate compositions is in turn determined by the structuring mechanism of NCO-containing fragment in the presence of PSO and a catalyst [107, 108]. There are two possible structuring mechanisms:

A:

$$\sim Ph-\overset{+}{N}H=C=O \quad \xrightarrow[\sim R\text{-}SH]{Ph'OH(OM)} \quad \sim Ph-\overset{H}{N}-\overset{\overset{\displaystyle O}{\|}}{C}-O-\underset{NR_3'}{Ph'} \quad \xrightarrow{\sim Ph\text{-}NCO}$$
$$\overset{-}{S}-R\sim$$

$$\sim Ph-\overset{\overset{\displaystyle O}{\|}}{\underset{\underset{\sim Ph-NH}{\overset{\displaystyle \|}{O=C}}}{N}}-\overset{NR_3'}{\underset{}{C}}-O-Ph' \quad \xrightarrow{\sim Ph\text{-}NCO} \quad \sim Ph-\overset{\overset{\displaystyle O}{\|}}{\underset{\underset{\underset{\overset{\displaystyle \|}{O}}{\sim Ph-N-C-N-Ph\sim}}{\overset{\displaystyle \|}{O=C}}}{N}}-\overset{NR_3'}{\underset{}{O}}\cdots Ph' \quad \xrightarrow{\sim Ph\text{-}NCO}$$

B:

$$\sim Ph-\overset{+}{N}H=C=O \quad \xrightarrow[\sim R\text{-}SH,\ H_2O]{\sim S\sim,\ NR_3'(OM)} \quad \sim Ph-\overset{H}{N}-\overset{\overset{\displaystyle O}{\|}}{C}-OH \quad \xrightarrow[CO_2\uparrow]{\sim Ph\text{-}NCO}$$
$$\overset{-}{S}-R\sim$$

$$\xrightarrow{\quad} \quad \sim Ph-\overset{H}{N}-\overset{\overset{\displaystyle O}{\|}}{C}-\overset{H}{N}-Ph\sim \quad \xrightarrow{\sim Ph\text{-}NCO} \quad \sim Ph-\overset{}{N}-\overset{\overset{\displaystyle O}{\|}}{C}-\overset{}{N}-Ph\sim$$
$$\underset{\sim Ph-NH \quad HN-Ph\sim}{O=C \quad O \quad C=O}$$

etc. Until the break of all the kinetic chains transferring active centers and the decomposition of the next charge transfer complex. The last process can also produce protons due to exhaustion of isocyanate groups.

The specific influence of MB (Ph'–OH, Ph'–NR$_3$) is in its ability to partially oxidize PSO in case of the next charge transfer complex decomposition. Thiourethane bonds may form here [109]:

$$\sim Ph-\overset{+}{N}H=C=O \quad \xrightarrow[\sim R\text{-}SH]{NR_3'(OM)} \quad \sim Ph-\overset{H}{N}-\overset{\overset{\displaystyle O}{\|}}{C}-OH \quad \rightleftharpoons^{NR_3'}$$
$$\overset{-}{S}-R\sim$$

$$\rightleftharpoons \quad \sim Ph\text{-}NH\text{-}\overset{\bullet}{C}=O \ + \ HO\text{-}N^+ \!{}^\bullet R'_3 \ ;$$

$$\sim Ph\text{-}NH\text{-}\overset{\bullet}{C}=O \xrightarrow{-R\text{-}SH} \sim Ph\text{-}NH\text{-}\underset{\underset{O}{|}}{\overset{\bullet}{C}H} \ + \ \sim R\text{-}S^\bullet \xrightarrow{\cdot\text{-}R\text{-}S^\bullet} \sim R\text{-}S\text{-}S\text{-}R\sim$$
$$\overset{\displaystyle \downarrow\phantom{xxxxxxxx}}{\phantom{x}\longrightarrow \ \sim Ph\text{-}NH_2 \ + \ CO\uparrow}$$

Such mechanisms seem to require more as the urethane formation reaction has been determined to occur mainly in the presence of tertiary amines [110] and has a substantially higher rate than urethane formation reaction, and copolymer structures finally form. Interpenetrating polymer network (IPN) principle can be applied to such systems for production of sealants only if a specific curing agent is used for each oligomer [111].

Phenol Mannich bases (and tertiary amines in general) are known to be catalysts for isocyanate groups isomerization (trimerization) and proton-mediated branching [64, 87, 102, 110, 112]:

$$3 \sim Ph-\overset{+}{N}H=C=O \quad \xrightarrow{\text{Ph}'-\text{OH(OM)}} \quad$$

(with $\overset{-}{S}-R\sim$ below the left structure)

If mercaptan is added, isocyanate prefers to form thiourethane, than enter into trimerization cycling [105].

The degree of conversion of reactive oligomers to thiourethane copolymer increases with PSO's functionality. More polar PSOs form more effective thiourethane polymer chain, when interact with isocyanates. Oligomers with similar polarity form more ordered polymer network structures [105].

Isocyanates have been made of urethane prepolymers with isocyanate end groups, in particular: diene urethane prepolymers (UP) (SKUDF-2), FP-65 [63, 105], as well as ester prepolymers oligoethylenediethylene-glycoladipinate (OEDGA)], ethylenebutyleneadipinate (SKUFE-4) [63]. The curing can be effectively carried out without heating with the use of amine type catalysts or phenolic Mannich bases [61, 64, 109]. Catalysts were introduced with volatile solvents and (or) plasticizers (dibutylphthalate, chloroparaffin, oligoesteracrylates) to improve their miscibility with oligomers.

Thiourethane can be made of polyesterpolysulfide (the product of interaction of polypropyleneglycol, epichlorhydrin, tin chloride penta hydrate and sodium disulfide) and polymethylenephenyl isocyanate in the presence of tertiary amine, as well as filler (titanium dioxide) and a plasticizer (dioctylphthalate) [113].

Thiourethane vulcanizates have, as a rule, poor physical and mechanical properties, if they are made of monomer diisocyanates. Therefore, it is recommended to get PSO adducts with the extent of diisocyanate and then add some hydroxyl-containing compounds to induce interaction with end groups of adducts. It is also proposed to make a prepolymer by mixing of PSO with 2,4-toluylenediisocyanate at 75–85°C and subsequent addition of diol at the PSO: TDI :diol ratio of 1:1.5:0.5. The first stage is prepolymer formation. Than it interacts with diol and turns into high-molecular weight polyurethane [114]. Diisocyanate-PSO adduct can also be produced with PSO taken in excess and then copolymerized with disocyanate until their ratio reaches SH: NCO=1: (0.9–1.1) [115]. This is the method to synthesize thiourethane compositions with optical properties.

There are studies concerning influence of OEDGA-50's molecular weight, oligothiol-prepolymer ratio at the preparation of thiourethane sealing composition, as well as oligothiol nature (PSO and TPM-2), on basic physico-mechanical properties of vulcanizates [63]. It has been determined, that oligoester with the molecular weight of M=2000÷2500 is required to make sealants with good deformation and strength properties, provided a liquid thiokol with M=1500÷2000 or TPM-2 polymer are used. If it is required to get PSO-based materials with higher relative elongation (500,650%), it is recommended to use oligoesters with higher molecular weight (M=3000÷3500), because it has been studied how oligoisocyanate's molecular weight and the quantity of MB and "Agidol AF-2" influence properties of thiourethane compositions.

Another study concerns influence of filler's type and quantity on properties of thiourethane compositions based on "SKUDF-2" × PSO [105]. The relative elongation has been determined to decrease gradually with the increase of filler content (carbon black). The conventional strength is maximal at some filler content, which depends on polarity of initial oligomers.

There is a research work concerning law patterns of isocyanate-oligothiol interaction, where effective catalysts has been found and thiourethane formation mechanisms have been proposed. There is research on estimation of MW influence for oligoester-based oligoisocyanates, as well as the oligothiol/prepolymer ratio, MB catalyst and "Agidol AF-2" content [63] on the properties of thiourethane compositions.

There is an interesting opportunity of getting curable compositions via formation of interpenetrating network-type structures, provided prepolymers with isocyanate and groups are used with PSOs. For example, two-component sealants can be made by introduction of a curing agent for PSO into urethane prepolymer (first component) and a curing agent into PSO (second component) or a curing catalyst for urethane prepolymer. Such an approach enables getting sealants with higher curing rate, improved strength and adhesion properties and increased chemical, oil and gasoline resistance [106, 107].

One of possible ways to modify PSO, being of an undisputable practical interest, is the change of SH-groups to isocyanate, epoxy or other groups in post-reactions [3].
Isocyanate

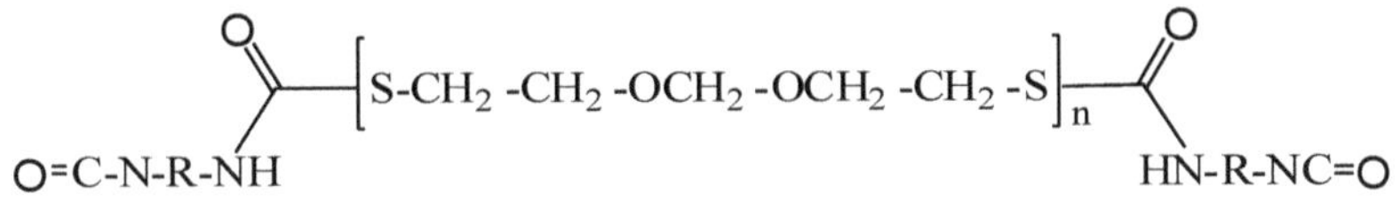

$$O=C-N-R-NH \qquad [S-CH_2-CH_2-OCH_2-OCH_2-CH_2-S]_n \qquad HN-R-NC=O$$

where R is the rest of diisocyanate.
Epoxy

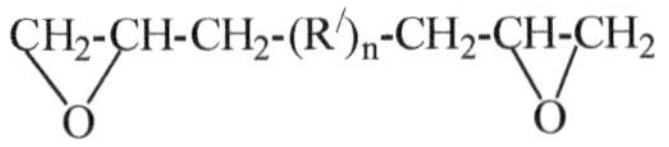

$$CH_2-CH-CH_2-(R')_n-CH_2-CH-CH_2 \qquad CH_2-CH-CH_2-O-R-(R')_6-R-O-CH_2-CH-CH_2$$

ELP-3 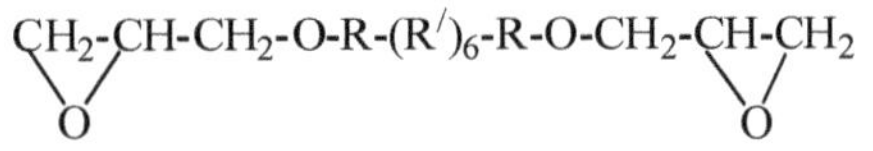                                   ELP-612

where R is the rest of bisphenol-A epoxy resin (type EEW 180).

R¢:-S-CH$_2$-CH$_2$O-CH$_2$O-CH$_2$-CH$_2$-S-

Interaction of end SH-groups of liquid thiokols with acrylate end groups of oligomers makes a series of hybrid sealants marked ZL-2244 and ZL-1866. They are characterized by good physico-mechanical properties, high elasticity, low water swelling and small water permeability:

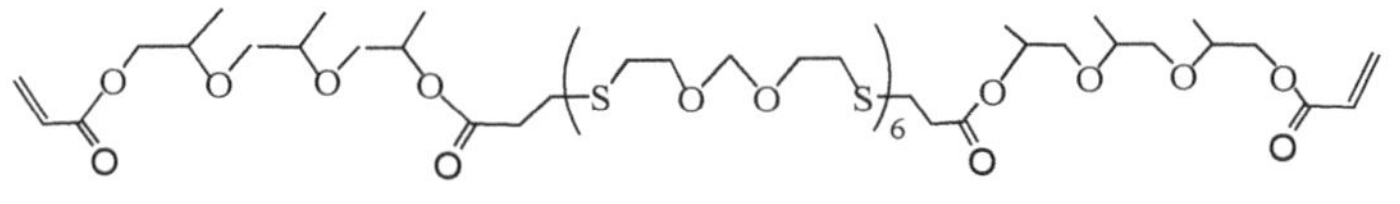

ZL - 2244

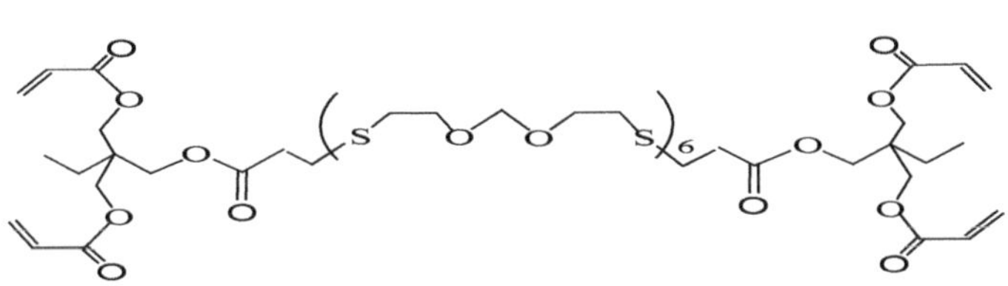

ZL - 1866

These examples do not limit the family of reactive oligomers used for modification of oligothiols. Modification of thiokol sealants increases, as a rule, their technological, strength and adhesive properties. Therefore, they seem to be prospective for solving the problem of broadening the production yield and quality improvement.

## 2.4   FILLERS FOR THIOKOL SEALANTS

The properties of PSO-based sealants mainly depend not only on the structure of oligomer, vulcanizing agent type and modifiers, but on the type and quantity of added fillers, plasticizers and other target additives as well.

The most common strengthening fillers for liquid thiokol sealants are carbon black of various grades, titanium dioxide, silicon dioxide, precipitated chalk and zinc sulfide. The group of non-strengthening fillers is also represented by neutral aluminum oxide, chalk, talk, mica, etc. Figure 2.1 demonstrates the influence of solvent nature of deformation and strength properties of thiokol sealants [1].

Two criteria are used for selection: the size of filler particles (dispersity) and water extract's pH. PSO vulcanization is carried out easier and faster in alkaline medium, when Thiokol's pH is within 6,8, Therefore, neutral fillers are recommended here [1].

Addition of carbon black to thiokol sealants increases module, strength, oil and gasoline resistance and other properties many times. The mechanism of such an increase phenomenon is similar to that of rubbers, made of non-crystallizing caouthoucs and mainly consists in formation of chain-like structures at certain critical carbon black concentration [118]. Fillers with small pH, such as kaolin, slow down liquid thiokol's vulcanization and decrease resistance of such sealants to increased temperatures. This effect can be caused by thiokol destruction via acetate bonds [1, 119].

The optimal carbon black (P-803) concentration in thiokol sealants is 30–35 mass parts, depending on technology and physico-chemical properties.

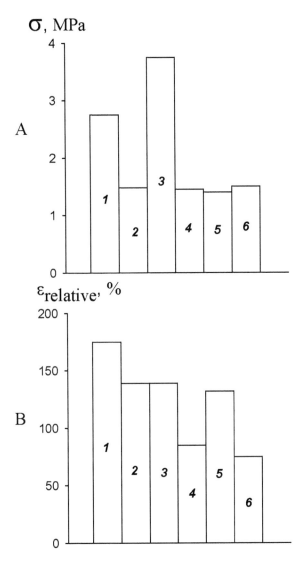

**FIGURE 2.1**    The influence of solvent nature on the properties of thiokol sealants. Filler content is 30 mass parts, manganese dioxide is used as a curing agent. 1 – Carbon black P-803; 2 – TiO$_2$; 3 – Carbon white BS-50; 4 – Nature chalk; 5–Lithomon; 6 – Kaolin.

The research [120] have determined, that carbon black of various grades, which mainly differ in their dispersity, forms the following row of strengthening effect intensity:

$$PM-15 > PM-70 > PM-50 > DG-100.$$

The increase of carbon black dispersity leads to the decrease of effective and chemical density of vulcanization network. This effect can be explained by impossibility of creating effective conditions for dispersing of filler into oligomer matrix, using existing equipment, first of all, for highly dispersed grades of strengthening fillers. This is typical for not only for liquid thiokol-based compositions, but for compositions containing curing liquid oligomers of other nature as well.

The increase of carbon black dispersity has been determined to result in a distinct change of molecular mobility, due to the change of a supermolecular structure. The increase of content and activity of a curing agent reduces the effect of carbon black dispersity on molecular mobility, due to the increase of vulcanization network density. EPR studies have shown, that liquid thiokol and carbon black interact only physically but not chemically [121, 122].

The research [123] is the study of carbon black influence on the average correlation time and molecular motion anisotropy in liquid thiokol. It has been found, that there can be a correlation between physico-mechanical properties of vulcanizates and the quantity of a curing agent and curing conditions.

There is the study [124] of how various fillers influence rheological properties, curing rate and properties of liquid thiokol-based sealants. The rate of liquid thiokol vulcanization by manganese dioxide reduces in the following row: Carbon black P-803 > Carbon white BS-120 > ZnO > Talc > Chalk MTD-2 > Titanium dioxide > Aerosil A-175 > Marshallite> Lime sweepings > Kaolin. It correlates with pH of fillers. According to DTA data, thermal destruction of sealants is initiated independently on the nature of filler in the range of 180-200°C, while visible mass loss occurs in the temperature range of 280–310°C. The content of carbon residues in sealants reduces in the following order: Talc > $TiO_2$ > ZnO > Kaolin > Marshallite > Carbon black P-803. It has been shown, that there is the following row of resistance to various media: Distilled water–Lime sweepings < Marshallite < Kaolin < Chalk MTD-2 < ZnO < $TiO_2$ < Talc < Aerosil A-175 < Carbon white BS-120 < Carbon black P-803; 5% sulfuric acid solution- Lime sweepings < Kaolin < ZnO < Chalk MTD-2 < Marshallite < $TiO_2$ < Talc < Aerosil A-175 < Carbon white BS-120 < Carbon black P-803; 5% sodium hydroxide solution–Kaolin ~ Chalk MTD-2 ~ Marshallite < Talc < ZnO < $TiO_2$ < Aerosil A-175 < Carbon black P-803 < Carbon white BS-120 < Lime sweepings. The increase of sulfuric acid concentration results in up to 10% and deeper loss of protective properties.

Chalk is widely used for production of sealants, despite its being non-strengthening filler, because of a wide feedstock base, relative cheapness and modification opportunities. Chalk is used as filler for PSO sealants together with plasticizers, thixotropic and adhesive additives. It makes it possible to produce various types of sealants having satisfactory strength and good elastic properties. These sealants are widely used for sealing of glass packets and interpanel joints [2].

Chalk blends with special modifiers are used to increase the quality of sealants in recent years. They often provide sealing compositions with thixotropic properties, as well as increase hydrophobicity of sealants and their adhesion to various substrates. There is also a chalk, treated with various surfactants. The type, concentration and addition method of surfactants influence rheological properties and stability of sealing pastes at storage [1].

Natural chalk grades are widely used for production of thiokol sealants for glass packets in recent years. These are highly dispersed grades of natural, chemically precipitated and hydrophobic chalk with particles up to 5 micrometers and concentration of 150 mass parts and more.

Highly dispersed mineral filler, which is mainly lamelli form calcium carbonate, has been determined to be effective [125]. Its introduction into two-component thiokol sealants provides them with better fluidity, thixotropicity and physico-mechanical properties, comparing precipitated chalk.

The study [126] summarizes the influence of a wide range of mineral fillers (such as natural and precipitated chalk and various types of clay) on a viscosity, thixotropicity and price of thiokol sealants. A mineral filler Polcarb, S., which is finely grinded natural chalk, exerts the most effective influence on the increase of thixotropicity.

The study [63, 105] concerns the influence of filler's type and quantity on properties of thiourethane compositions (up to 40 mass parts of carbon black and up to 90 mass parts of chalk per 100 mass parts of oligomer). The increase of filler content has been determined to result in the increase of strength. Comparative analysis indicates correlation between the strength of sealants and the degree of polarity of initial oligomers (sealant components).

There was a study of how a combination of low dispersed and highly dispersed sealants of the same nature (P-514 and P-803 grade carbon black, anatase and rutile titanium dioxide) and a combination of fillers varied in nature and dispersity (chalk and titanium dioxide), effects properties of thiokol sealants. The use of such combinations has been determined to reduce viscosity that is explained by the increase of unbound oligomer in a composition, due to denser packing of filler particles "free volume" effect. At the same time, elasticity of cured compositions increases [127].

Another fillers for thiokol sealants are finely dispersed sand or slag, glass-ceramics, andesite flour, asbestos. Their application helps to obtain cheaper materials and, in some cases, to increase dielectric properties, moisture, impact and heat resistance of sealants [1].

## 2.5   PLASTICIZERS FOR THIOKOL SEALANTS

Plasticizers are widely used to regulate rheological properties of uncured compositions, to increase deformation properties and to cheapen sealants. However, they decrease elongation rate, tear resistance and adhesive properties of sealants and can make them thermoplastic at increased temperatures [1].

Plasticizers favor distribution of powder-like vulcanizing agents and are used to produce so-called vulcanizing pastes, whose application substantially improves set of properties of thiokol sealants.

Plasticizers can exert influence on PSO's curing rate. The more polarity and pH value of as plasticizer is, the faster thiokol sealant is cured.

The most widely used plasticizers are compounds, which are well compatible with thiokol, such as chloropropane of various grades, dibutylphthalate, benzylbutylphthalate benzyloctylphthalate, diphenyl chloride, etc. [1, 2, 11].

To choose a plasticizer for a specific composition, it is necessary to consider the nature of PSO, its polarity (compatibility with PSO), volatility and other properties, processing and operation conditions of sealants as well as the application range of a composition. For example, benzylbutylphthalate or similar plasticizers are recommended for thiokol sealants, applied for production of glass packets, because they have antifog properties and a minimal volatility [2].

There are studies, concerning influence of various plasticizers on their compatibility with U30 M thiokol, vulcanization rate, rheological, physico-mechanical and dielectric properties, as well as on thiokol sealant's vulcanization network structure. The following plasticizers were used: HP-30, HP-52, HP-470, flotation agent-oxale and a complex plasticizer PL-105, which is the mixture of dioxane alcohols, their high-boiling esters and chloroparaffin [128, 129]. Studied plasticizers form the following row of compatibility with liquid thiokol: "Netoxol" < HP-30 < HP-470 < HP-52 < DBF < PL-105 < Flotation agent. Vulcanization rate reduces with the increase of plasticizer content. The effective crosslink density is maximal, when its content is 7.5 mass parts, and this value is independent on plasticizer's nature. The maximum strength is observed for HP-470 and HP-52 sealants. The diffusion of aggressive media in sealants containing studied plasticizers increases at their content above 10 mass parts. The optimal range of properties proves to be provided by flotation agent-oxale and PL-105-grade plasticizer.

The following plasticizers can substantially decrease the cost of thiokol sealants, meant for road carpets: mineral rubber, bitumen, tars, liquid factices, anthracene oil, petroleum and coal-tar resins and tar oils, added from 5 to 50 mass parts per 100 mass parts of an oligomer [1, 2, 130, 131].

## 2.6   ADHESIVE AND MODIFYING ADDITIVES

Thiokol sealants are known to be unsatisfactory adhesive without modifying additives, which are usually reactive. Therefore, thiokol sealants are blended with adhesive additives, because such sealants are mainly used for coating, and one of their principal characteristics is adhesion to various substrates to provide durability. Such additives are usually reactive compounds, which can react chemically both with PSOs and a surface being sealed. In addition to improvement of adhesive properties, these compounds can influence vulcanization rate, vulcanization network parameters, physico-mechanical properties, behavior in various aging conditions, because they participate in vulcanization. This fact is necessary to be considered during application-oriented design and selection of sealants.

There are many compounds, offered as adhesive additives. However, the most effective are epoxy resins, alkyl-phenol formaldehyde resins, monomer and oligomeric acrylates and functionalized alkoxysilanes [1, 2, 11].

Epoxy Diane resins ED-20 and E-40 at the quantity of 5–13 mass parts are the most widespread adhesion modifiers for thiokol sealants used in aircraft, machinery and shipbuilding and applied to duralumin, steel or concrete [75]. Epoxy resins increases adhesive properties of thiokol sealants through their interaction with thiokol and formation of hydroxyl groups, which actively form adhesive bonds with substrate. The chemistry of epoxy resin-liquid thiokol interaction is described in Chapter 1.5.2. Resin does not fully interact with thiokol under curing conditions. It has been determined, that only 2.5–3 mass parts of epoxy resin per added 5–10 mass parts bind chemically with thiokol. The rest of resin stays unbound inside bulk sealant and acts as a plasticizer for it, substantially reducing its strength properties [82, 132].

It is known, that PSOs can be modified by alkyl phenol formal dehyde resins (PFR). PFRs can be added at the stage of PSO drying during its synthesis [133,134]. Thiol end groups react easily with methyl groups of PFR and form simple thioether bonds. Addition of PFR up to 30% increases adhesion of sealants to various substrates:

$$\sim\text{R-SH} + \text{HO-CH}_2\text{-Ph(OH)-R}'\sim \longrightarrow \sim\text{R-S-CH}_2\text{-Ph(OH)-R}'\sim + \text{H}_2\text{O}$$

Sealant properties weakly depend on the method of resin addition, Therefore, the simplest variant is used in industry: PFR is added in a form of alcohol solution (VITEF-grade sealant) [135].

Methylmetacrylate and his derivatives, such as oligoesteracrylates, are of the widest practical interest among unsaturated reactive modifiers.

Acrylates can modify oxidation-curable thiokol sealants by a chemical interaction with thiokol. The mechanism of acrylate–liquid thiokol interaction is given below:

$$\sim RS^* + CH_2=\underset{\underset{CH_3}{|}}{C}-R^/\sim \longrightarrow \sim RS-CH_2-\underset{\underset{CH_3}{|}}{C}^*-R^/\sim$$

$$\sim RS-CH_2-\underset{\underset{CH_3}{|}}{\overset{\overset{R^/}{|}}{C}}^* + {}^*SR\sim \longrightarrow \sim RS-CH_2-\underset{\underset{CH_3}{|}}{\overset{\overset{R^/}{|}}{C}}-SR\sim$$

$$\sim RS-CH_2-\underset{\underset{CH_3}{|}}{\overset{\overset{R^/}{|}}{C}}^* + \sim RS-SR\sim \longrightarrow \sim RS-CH_2-\underset{\underset{CH_3}{|}}{\overset{\overset{R^/}{|}}{C}}-SR\sim + \sim RS^*$$

–RS· radical can form via the removal of hydrogen atom from SH-group or by the homolytic decomposition of disulfide bonds in liquid thiokol [1, 136].

Interaction is the most effective when liquid thiokols are cured by sodium bichromate or manganese dioxide. Introduction of 5-10 mass parts of acrylates substantially reduces viscosity of compositions and improves physico-chemical and adhesion properties as well. Thus, composition with filling properties can be prepared by this method [137–139]. Oligoester-acrylates form the following row of influence on physico-mechanical and strength properties of thiokol sealants [140]:

$$7-1 > 7-20 > TMGPh-11 > TGM-3$$

Glycidylmetacrylate, styrene solutions of PN-1 and PN-9119-grade [124] polyester resins and polymeric complexes of acrylic or methacrylic acid with e-caprolactam have been found to improve adhesive and strength properties of Thiokol sealants. The modifying effect for polymer complexes is maximal, when aqueous solutions of polymer complex and sodium bichromate are used [141].

One of the most effective ways to increase adhesive properties is modification of PSO by silicon compounds. These compounds are supposed to have a functional group on the one side, which is able to react with SH-group (vinyl, epoxy group etc.). The other side is to contain alkoxy groups, being able to hydrolyze by aerial moisture forming reactive silanol groups, which actively form chemical bonds with fillers and such substrates as glass, metal, concrete, etc. [1, 3]:

$$R \left( \overset{H_2}{\underset{}{C}} \right)_3 - \overset{H_2}{\underset{}{Si}} \left( OR' \right)_3,$$

where $R - NH_2$, vinyl, epoxy, SH-group etc., $R'-C_2H_5$, $CH_3$, etc.

The interaction mechanism of such silanes with thiokol and substrate's hydroxyl groups is given below. Vinyltriethoxysilane is taken for example:

$$\sim R\text{-}S\text{-}CH_2\text{-}CH_2\text{-}(CH_2)_3\text{-}\underset{\underset{OC_2H_5}{|}}{\overset{\overset{OC_2H_5}{|}}{Si}}\text{-}OC_2H_5 + H_2O \longrightarrow \qquad (1)$$

$$\longrightarrow \sim R\text{-}S\text{-}CH_2\text{-}CH_2\text{-}(CH_2)_3\text{-}\underset{\underset{OC_2H_5}{|}}{\overset{\overset{OC_2H_5}{|}}{Si}}\text{-}OH + C_2H_5OH$$

$$\sim R\text{-}S\text{-}CH_2\text{-}CH_2\text{-}(CH_2)_3\text{-}\underset{\underset{OC_2H_5}{|}}{\overset{\overset{OC_2H_5}{|}}{Si}}\text{-}OH + HO\text{-}R'\!\sim \longrightarrow \qquad (2)$$

$$\longrightarrow \sim R\text{-}S\text{-}CH_2\text{-}CH_2\text{-}(CH_2)_3\text{-}\underset{\underset{OC_2H_5}{|}}{\overset{\overset{OC_2H_5}{|}}{Si}}\text{-}OR'\!\sim + H_2O \quad \text{и т. д.}$$

Such silanes are effective even at small concentrations, and their content in a composition is usually 0.1–1.5 mass parts.

Such sealants have been known to increase adhesive properties of thiokol sealants for 50 years [142, 143]. However, it were 1980s as the starting point of their wide usage as components of thiokol sealants to seal second (external) side of glass packets, which were extensively applied in construction.

Although silanes themselves are the most widespread adhesive additives in construction and industry (first of all, silanes with epoxy, vinyl and mercaptan groups), it is also possible to use such silanes in both a hydrolyzed form and as adducts, for example, with epoxy resins [4, 11, 144, 145]. Other proposed adhesive additives are silicicated amines [124].

## 2.7   VULCANIZATION PROMOTERS AND RETARDERS

Promoters and retarders are primarily meant to regulate the curing rate of thiokol sealants. Alkaline substances are curing promoters and acidic substances are curing retarders. There is a direct relation between curing rate and the basicity of medium (in other words, the content of diphenylguanidine), which is shown in the Fig. 2.2 [1].

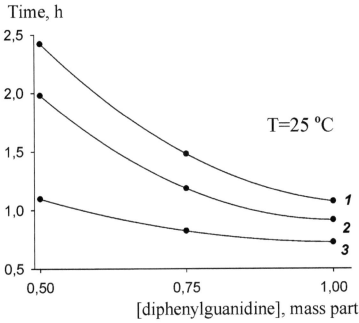

FIGURE 2.2    The dependence of viability of "U-30 M"-grade thiokol sealant on the content of vulcanization promoter and paste-like manganese dioxide: 1–5 mass parts; 2–7 mass parts; 3–9 mass parts, paste number 9.

Amines are the most popular substances for two-component sealants, cured by manganese or lead dioxides. Such compounds are mainly basic and promotion is achieved by a considerable polarization of SH-bonds by amine's nitrogen. There are many amino compounds described in scientific literature, which promote vulcanization of thiokols [1, 11, 24, 124]. However, the most popular in practice are diphenylguanidine, thiuram and sulfur. Amines are only promoters and do not influence network formation processes and the final properties of vulcanizates. But thiuram is not only oxidizing promoter, but a participant of interchain exchange reactions as well, thanks to constituent disulfide bonds. As a result, its excess can destruct thiokol's main chain. Sulfur can itself oxidize SH-groups of thiokol. Its small additives effectively promote curing of thiokol by metal dioxides.

Its content should not usually exceed 0,2 mass parts as its further increase will deteriorate adhesion and soften vulcanizate.

Effective promoters for liquid thiokol curing by sodium bichromate are water, ethanolamines, diphenylguanidine.

Active promoters for organic peroxides are ethanolamines and tertiary amines [1, 141].

Effective promoting agents for curing of liquid thiokols by calcium and zinc dioxides are water, sulfur, amines and various sulfurous promoters [1, 2, 11, 48, 49]. Acidic compounds [52, 53] and sulfur-and-nitrogen-containing compounds [146] are effective promoters for curing of liquid thiokols by zinc oxide.

$$(CaO_2 + H_2O \longrightarrow Ca(OH)_2 + O)$$

When tailored regulation of vulcanization rate is needed, vulcanization retarders are applied with promoters. Such substances are usually acidic. The most widespread vulcanization retarders are stearic, isostearic and oleic acids, added in the amount of 0.1–3.0 mass parts [1, 2]. Less retarding effect is demonstrated by α-branched monocarbonic acids $C_{21}$–$C_{27}$ and synthetic fatty acids $C_{10}$–$C_{20}$ [1, 147]. There is information about application of organic compounds, containing halogen atoms, such as chlorine and bromine [148, 149]. They effectively retard oxidation processes in the amount of 0, 5 mass parts. Organic acid anhydrides are also used, such as acetic, propionic, butyric, maleic, benzoic acids, etc. [150].

The most popular components of a vulcanizing composition are given in Table 2.2, depending on a vulcanizing agent type and the application range of thiokol sealants.

**TABLE 2.2**   Types of Vulcanizing Systems For Liquid Thiokols

| Vulcanizing agent | Promoter | Retarder | Application aspects and ranges |
| --- | --- | --- | --- |
| $PbO_2$ | Sulfur, amines | Stearic acid, lead stearate | Widespread occurrence. Easily controlled curing. Construction sealants |
| $MnO_2$ | Inorganic bases, amines | Stearic and isostearic acids | Widespread occurrence. More thermally resistant sealants, than $PbO_2$ based ones. High thermal, UV and chemical resistance. Aircraft, glass packets, construction sealants |

**TABLE 2.2**   *(Continued)*

| Vulcanizing agent | Promoter | Retarder | Application aspects and ranges |
|---|---|---|---|
| $Na_2Cr_2O_7$ | Water, ethanolamines, diphenylguanidine, bases | Stearic acid, oleic acid and its esters | High thermal resistance and physico-mechanical properties. Aircraft, manufacturing engineering |
| $CaO_2$ $BaO_2$ | Water, amines, bases | Molecular sieves | Single-component sealants of white color |
| $ZnO_2$ | Amines, tetramethylthiuramdisulfide | Sulfur, molecular sieves | Single-and-two-component sealants of white color. Construction, glass covers |
| Sodium perborate | Water, amines, bases | Molecular sieves | UV-resistance. Single-and-two-component sealants of white color. Construction, glass covers |

## 2.8   THIXOTROPIC ADDITIVES

The most widespread thixotropic additives for PSO are aerosils, carbon white and bentonites [1, 2, 11]. It is also recommended to use hydrogenated castor oil or aerosol combination with compounds, containing hydroxyl groups [1, 11]. Aerosils with the specific surface area of 130–200 m²/g are mainly used, due to problems, appearing during bulk distribution and mixing. Aerosil content can vary with sealant's purpose and thixotropicity requirements and reach 5.0 or more mass parts per 100 mass parts of thiokol. However, research on aerosol influence on the properties of thiokol sealants is extremely scarce. There is fragmentary and scanty information about other substances, which can provide thiokol sealants with thixotropic properties.

## 2.9   SINGLE-COMPONENT LIQUID THIOKOL SEALANTS

There are many patents for making of single-component thiokol sealants [1, 11]. However, unlike single-component silicon sealants, they are not widely used, although they are commercially produced. It is first of all due to insufficient curing rate of thiokol sealants because of their low gas permeability. Therefore, single-component thiokol sealants are not recommended to use in construction for sealing interpanel junctions, because seams, made of such sealants, will not achieve a minimal critical strength. Therefore, it will break due to alternating deformations, caused by daily temperature variations. In addition, single-component thiokol sealants cannot be used to seal glass packets, if fast curing rate is required (sealant viability of 30–60 min). It is impossible to use them in aircraft or shipbuilding for same reasons. There are several approaches to making single-component thiokol sealants. They can all be reduced to activation of constituent vulcanizing agents or promoters by air moisture (curing by oxygen):

1. Application of dried manganese dioxide as an oxidizer in combination with classic driers (ceolites, bentonites, MgO, BAO, etc). Curing is initiated at the contact with air moisture and is limited by its diffusion in bulk sealant.

2. Application of latent curing agents or promoters can be based on two principles. The first one is capsulation of a curing agent. It is the most trivial method. Curing is initiated by the mechanical decomposition of a capsule. The second method is promoter's isolation, for example, in ceolites. A promoter is extracted from ceolite after the contact with air moisture. So called chemically "blocked" promoters can also be used. They are extracted in active form after interaction with air moisture.

3. Application of calcium or zinc dioxides, sodium perborate, anhydrous sodium bichromate. Curing is also activated by water, and the curing rate is limited by its diffusion into sealant's bulk.

4. Post-reaction, resulting in formation of so-called "hybrid" thiokol with functional groups being able to carry out effective curing in a single-component product without a curing agent.

5. Oxidation by aerial oxygen. A sealant is blended by siccative compounds, which activate oxidation processes. In contrast to methods with other curing agents, this method is similar to the technology, involving manganese dioxide, as it allows making sealants with maximum oil and gasoline resistance. Curing promoters for single-component sealants can be both specific and ordinary compounds, which are used for two-component compositions. Publications [1, 2, 11] provide the explicit description of compositions and properties of single-component sealants, based on liquid thiokol as well as related patents.

## 2.10 PROPERTIES, GRADES AND APPLICATION OF SEALANTS, BASED ON POLYSULFIDE OLIGOMERS

Liquid thiokol vulcanizates have the following range of properties, stipulated by high sulfur content: high resistance to various aggressive media; fuel, solvents, atmospheric aging in a temperature range of 55 to +130°C, oxygen, ozone, UV and IR radiation; high resistance to thermal and radiation aging, high moisture-, vapor- and gas permeability; satisfactory physico-mechanical and dielectric properties; small adhesion to the majority of substrates, poor wear and tear resistance.

Liquid thiokols are used to produce sealing materials for application in aircraft industry for sealing fuel tanks and airplane cabins, fuselage sealants; air, oil and fuel lines as well as in shipbuilding for caulking of decks, case constructions with various riveted and bolt joints, as well as for conservation of ships, in automobile and tractor industry, electrical technology and radio electronics. This application range is stipulated by their processing and curing capabilities at ordinary temperatures, which occurs and almost without shrinkage, extreme atmospheric, solvent and water resistance in a wide temperature range [10, 11, 75, 151–154].

Liquid thiokol sealants are widely used in construction for sealing interpanel joints, large-block construction, production of glass packets, as well as in road construction for sealing temperature, deformation and shrinkage seams in concrete and asphalt carpets, landing strips of aerodromes, etc.

A series of thiokol sealants has been worked-out for aircraft, mechanical engineering and shipbuilding. Their properties are summarized in Tables 2.3 and 2.4 [11, 75, 151].

U30 M, UT31, UT32, U30 MES-5, U30 MES-10-grade sealants are designed for sealing bolt, riveted and other metal joints in aircraft, shipbuilding, transport and mechanical engineering industries, which are subjected to air and fuel influence. They are highly resistant to aging in solvents. U30 M and UT31-grade sealants are used with de sealants are used with adhesive sub layers based on chloroprene or epoxy-thiokol glues [151]. UT34-grade sealant if designed for sealing various metallic joints subjected to the effect of air or fuels as well as for sealing plug couplers and various devices. It has filling properties and can be applied by putty knife, syringe, brush or simply filled. 51UT36-grade sealant is thixotropic and designed for sealing various metal structures for outdoor use. 51UT37-

grade sealant is thixotropic too and is designed for sealing metal structures, used on air or in sea or river water. 51UT38-grade sealant is used in highway engineering and is designed for waterproofing of temperature and deformation seams of highways, landing strips, etc. It is durable at usage bothoutdoors and in water. "VITEF"-grade sealant is designed for surface or seam sealing of metal joints and glass cover elements subjected to the influence of air or fuels. All the above listed sealants are not supposed to be used for sealing copper, brass or silver joints. U30 M, UT31, UT32, UT34, U30 MES-5 and U30 MES-10p grade sealants are cured by manganese dioxide, 51UT36, 51UT38 and "VITEF" by sodium bichromate and 51UT37 – by lead dioxide. All the sealants are three-component, excluding 51UT38.

Thiokol sealants are highly oil- gasoline- and water resistant. They can be durably used in these media in a broad temperature range (Table 2.4). When they are in contact with fuel at increased temperatures, their strength increase, so do relative and residual elongation. Thiokol sealants are highly resistant to alcohols, ethers, glycols, gasolines, diluted acids and alkalis and are less resistant to acetone, ethyl acetate, methyl ethyl ketone, benzene, dichloroethane and other polar solvents (Table 2.5).

Thiokol sealants are highly resistant to ozone (Table 2.6).Thiokol sealants have small shrinkage (the value of shrinkage factor is 0.4–0.6%).

The degree of elastic recovery of thiokol sealants strongly depends on a vulcanizing agent type. Its value is 80–95% for manganese dioxide, 70–80% for lead dioxide, 70–85% for dicumyl peroxide, and 50–70% for calcium and zinc dioxides.

Thiokol sealants are well resistant to radiation. The relative elongation at radiation dose of $5.16 \times 10^4$ Coulomb/kg does not drop below 100%, their strength stays unchanged and adhesion at peeling to steel or duralumin increases 1.5–2.5 times. Studies, carried out by the authors, have confirmed high radiation resistance of thiokol sealants. For example, deformation and strength properties of U-30 M and U-30 MES5-grade sealants stay almost unchanged after radiation of (PB-1200 radiation plant) up to 500 kilograms.

Liquid thiokol vulcanizates have high water and gas permeability. Steam permeability is within 1–6 $g/m^2$ per 24 h (DIN 53122), depending on a composition of thiokol sealant, while argon permeability is within 40–70 $mL/m^2$ sm bar [2].

**TABLE 2.3** Basic Properties of Thiokol Sealants, Used in Aircraft and Mechanical Engineering

| Grade | Color | Consistency | Viability, hours | Density kg/m³ | Breaking stress, MPa | Relative elongation at rupture, % | Residual elongation, % | Shore hardness A. conv. units | Tear resistance, kN/m | Resistance of peeling from duralumin D-16AT, kN/m | The stress at 100%-elongation, MPa | Brittle point, °C | Working temperature range °C |
|---|---|---|---|---|---|---|---|---|---|---|---|---|---|
| 1 | 2 | 3 | 4 | 5 | 6 | 7 | 8 | 9 | 10 | 11 | 12 | 13 | 14 |
| U-30M | Black | Viscous-flow | 2-9 | 1400 | 2.5-4.0 | 150-300 | 0-8 | 50-65 | 1,175-1,470 | 1,71-2,94 (with sublayer) | 2,0-3,0 | From -40 to -50 | -60 ÷ +130 |
| U-30MES-5 | | | 2-10 | 1400 | 1.5-3.0 | 180-350 | 0-12 | 40-55 | 0,980-1,175 | 1,47-2,94 | 1,0-1,2 | From -40 to -45 | -60 ÷ +130 |
| U-30MES-10 | | | 2-10 | 1400 | 1.2-2.0 | 220-500 | 2-20 | 30-45 | 0,780-0,980 | 1,47-2,94 (from steel) | 0,8-1,0 | From -40 to -45 | -60 ÷ +130 |
| UT-31 | Light gray | Paste-like | 2-9 | 1950 | 2.0-3.5 | 175-300 | 0-10 | 50-65 | 1,175-1,470 | 1,71-2,94 (with sublayer) | 1,0-1,5 | From -40 to -50 | -60 ÷ +130 |
| UT-32 | Gray | | 2-8 | 1700 | 1.5-2.5 | 200-600 | 0-10 | 35-50 | 0,785-1,175 | 1,71-2,94 | 0,5-0,8 | From -40 to -45 | -60 ÷ +130 |
| UT-34 | | Viscous-flow | 3-25 | 1550 | 1.0-1.5 | 150-300 | 2-20 | 25-40 | 0,590-0,785 | 1,71-2,94 | 0,4-0,5 | From -40 to -45 | -60 ÷ +130 |
| 51-UT-36A | Dark-gray | Ointment-like | 1-3 | 1450 | 4.0-5.5 | 200-350 | 0-10 | 65-80 | 1,275-1,665 | - | 2,5-3,5 | From -40 to -45 | -60 ÷ +130 |
| 51-UT-36b | | | 1-5 | 1450 | 2.5-4.0 | 200-400 | 0-10 | 50-65 | 0,980-1,275 | 0,98-1,96 | - | From -38 to -43 | -60 ÷ +130 |
| 51-UT-37 | Brown | Paste-like, thixotropy | 1-4 | 1400 | 2.0-3.0 | 200-300 | 0-15 | - | - | 1,96-3,92 | - | From -30 to -35 | -40 ÷ +100 |
| 51-UT-38A | Black | Viscous-flow, mobile | 2-6 | 1350 | 2.0-3.0 | 250-400 | 5-10 | 40-50 | 0,980-1,175 | - | 1,0-1,2 | From -30 to -35 | -40 ÷ -70 |
| 51-UT-38B | | The same | 2-10 | 1300 | 1.5-1.8 | 300-600 | 10-20 | 30-40 | - | - | 0,4-0,8 | From -30 to -35 | |
| 51-UT-38G | | | 4-12 | 1300 | 0,8-2,0 | 300-600 | ≤80 | 30-40 | - | - | 0,2-0,8 | From -25 to -30 | |
| VITEF-1 | Beige | Paste-like | 2-8 | 1500 | ≥1.5 | ≥160 | ≤8 | 30-45 | - | ≥1,960 | - | - | -60 ÷ +130 |
| VITEF-2 | Light gray | Viscous-flow | ≤8 | 1470 | ≥1.6 | ≥160 | ≤8 | 30-45 | - | ≥1,960 | - | - | -60 ÷ +130 |

**TABLE 2.4** Heat Resistance and Dielectric Properties of Thiokol Sealants

| Medium | Temperature, °C | Service life, years | U-30M | U-30MES5 | U-30MES10 | UT-31 | UT-32 | UT-34 | 51-UT 36 | 51-UT 37 | VITEF 1,2 |
|---|---|---|---|---|---|---|---|---|---|---|---|
| Air | 25 | Years | 30-40 | 15-20 | 15-20 | 30-40 | 15-20 | 15-20 | 15-20 | 15-20 | 15-20 |
| | 50 | Years | 8-10 | 5-7 | 5-7 | 8-10 | 5-7 | 5-7 | 5-7 | 5-7 | 5-7 |
| | 70 | Years | 3-3,5 | 1-1,5 | 1-1,5 | 3,5 | 1-1,5 | 1-1,5 | 1-1,5 | 1-1,5 | 1-1,5 |
| | 100 | Hours | 1500-2000 | 1200-1500 | 1200-1500 | 1500-2000 | 1200-1500 | 1200-1500 | 1200-1500 | 1200-1500 | 1200-1500 |
| | 130 | Hours | 300-500 | 200-300 | 300-300 | 300-500 | 200-300 | 300-300 | 300-300 | 300-300 | 300-300 |
| | 150 | Hours | 40-70 | 30-50 | 20-50 | 40-70 | - | - | - | - | - |
| Sea water | -4 - +35 | Hours | 70000 | 70000 | 70000 | - | 70000 | - | - | 70000 | - |
| T-1-grade fuel | 20 | Hours | - | 134000 | - | - | - | - | - | - | - |
| | 50 | Hours | - | 85000 | - | - | - | - | - | - | - |
| | 130 | Hours | - | - | - | - | - | - | - | - | 50 |
| Characteristics | | | | | | | | | | | |
| -Specific surface resistance, ohm | | | $1 \cdot 10^8 - 5 \cdot 10^9$ | $1 \cdot 10^8 - 5 \cdot 10^9$ | $1 \cdot 10^8 - 5 \cdot 10^9$ | $1 \cdot 10^9 - 5 \cdot 10^{11}$ | $1 \cdot 10^{10} - 1 \cdot 10^{11}$ | $1 \cdot 10^{10} - 5 \cdot 10^{11}$ | $1 \cdot 10^{10} - 5 \cdot 10^{11}$ | $1 \cdot 10^{10} - 5 \cdot 10^{11}$ | $1 \cdot 10^{10} - 5 \cdot 10^{11}$ |
| -Specific volume resistance, ohm·m | | | $1 \cdot 10^6 - 1 \cdot 10^7$ | $1 \cdot 10^6 - 1 \cdot 10^7$ | $1 \cdot 10^6 - 1 \cdot 10^7$ | $1 \cdot 10^8 - 1 \cdot 10^9$ | $1 \cdot 10^8 - 1 \cdot 10^9$ | $1 \cdot 10^8 - 1 \cdot 10^9$ | $1 \cdot 10^8 - 1 \cdot 10^9$ | $1 \cdot 10^8 - 1 \cdot 10^9$ | $1 \cdot 10^8 - 1 \cdot 10^9$ |
| -Loss-angle tangent | | | - | - | - | 0,05-0,1 | 0,05-0,1 | 0,05-0,1 | 0,05-0,1 | 0,05-0,1 | 0,05-0,1 |
| -Dielectric constant | | | - | - | - | 8-10 | 8-10 | 8-10 | 8-10 | 8-10 | 8-10 |
| -Electric strength, kV/mm | | | - | - | - | 1,0 | 1,0 | 1,0 | 1,0 | 1,0 | 1,0 |

| | U-30 M | U-30 MES-10 |
|---|---|---|
| Linear expansion factor, $°C^{-1}$ | | |
| Above glass-transition temperature | - | $1.75*10^{-4}$ |
| Beyond glass-transition temperature | - | $0.50*10^{-4}$ |
| Heat conductivity, Watt/(m*°C) at 20°C | 0.430 | 0.362 |
| at 60°C | 0.454 | - |
| at 100°C | 0.467 | - |
| Thermal diffusivity, $m^2/c$ at 20°C | $3.72*10^{-11}$ | $3.19*10^{-11}$ |
| at 60°C | $3.55*10^{-11}$ | - |
| at 100°C | $0.56*10^{-11}$ | - |
| Specific heat capacity, kilojoules/(kg•°C) | 1.59 | 1.53 |

**TABLE 2.5**   The Resistance of Liquid Thiokol Vulcanizates of U-30 M and UT-31-Grades (250 h, 20°C) and LP-2-Grade Based Vulcanizates (720 h, 25°!) in Various Media

| Medium | U 30 M | UT 31 | Based on LP-2-grade thiokol |
|---|---|---|---|
| 1 | 2 | 3 | 4 |
| gasoline | (0.1) −0.9 | −0.3 ( 0.2–0.7) | – |
| kerosene T-1 | 0.2–0.3 3.4 | 3.2 (0.2–0.5) | – |
| Autol-18 | −0.2 | −0.1 | – |
| benzene | (60,70 ) 87.5 | (60,70 ) 95.7 | 195 |
| xylene | 24.5 | 21.1 | 39 |
| styrene | 103.8 | 99.5 | – |
| toluene | 47.2 | 47.0 | 95 |
| turpentine | 11.8 | 6.4 | – |
| machine oil СУ | −2.7 | −2.3 | – |
| transformer oil | −2.6 | −1.8 | – |
| chlorobenzene | 147.7 | 167.5 | 270 |
| carbon tetrachloride | 66.8 | 54.9 | 55 |
| butyleneglycol | −0.3 | −0.2 | – |
| ethylene glycol | −0.2 | 0.4 | −3 |
| butyl alcohol | 1.0 | −0.3 | −4 |
| glycerol | −0.8 | 0.8 | −2 |
| linseed oil | −0.2 | −2.7 | −4 |
| methyl alcohol | −0.6 | 0.4 | 6 |

**TABLE 2.5** *(Continued)*

| Medium | U 30 M | UT 31 | Based on LP-2-grade thiokol |
|---|---|---|---|
| ethyl alcohol | –0.8 | –0.3 | –5 |
| amyl acetate | 13.4 | 11.1 | – |
| dibutylphthalate | 17.1 | 16.0 | 30 |
| diethyl ether | 3.7 | 3.4 | – |
| tricresyl phosphate | 34.8 | 26.8 | – |
| nitrobenzene | >200 | >200 | – |
| acetone | 19.2 | 17.0 | 19 |
| methylethylketone | 28.8 | 29.1 | 56 |
| furfural | 143.8 | 139.9 | – |
| cyclohexanone | (40–50 ) | (100–130 ) | – |
| fresh water | (0.3–0.5 ) | (1–3 ) | 7– |
| sea water | (0.6–0.75) | (0.8–1.0) |  |
| nitric acid 10%, 20°C | nonpersistent | nonpersistent | nonpersistent |
| boric acid 2%, 60°C | persistent | persistent | – |
| formic acid 10%, 20°C | nonpersistent | nonpersistent | – |
| sulfuric acid 10%, 20°C<br>20%, 20°C<br>20%, 70°C | persistent<br>persistent<br>persistent | persistent (UT32)<br>persistent (UT32)<br>persistent (UT32) | 1<br>0<br>6 (50%) |
| hydrochloric acid 10%, 70°C | persistent | persistent (UT–32) | – |
| acetic acid 20%, 70°C | persistent | persistent | 12 |
| phosphoric acid 20%, 20°C<br>20%, 60°C | persistent<br>persistent | persistent non-<br>persistent | –<br>– |
| chromic acid 10%, 20°C | nonpersistent | nonpersistent | – |
| sodium hydroxide 40%, 70°C | persistent | persistent | – |

In brackets: the swelling degree in 24 h at 20°C. LP-2 thiokol sealant composition: thiokol-100, stearic acid-1, carbon black – SRF, lead dioxide curing paste.

**TABLE 2.6**   Resistance of U30 M and UT31-Grade Thiokol Sealants to Ozone*

| Ozone concentration, % | Deformation of a sample % | U 30 M | UT-31 |
|---|---|---|---|
| 0,01 | 10 | 15–25 | 15–25 |
| 0,01 | 20 | 6–12 | 6–12 |
| 0,1 | 10 | 4–5 | 4–5 |
| 0,1 | 20 | 1–1.5 | 1–1.5 |

*hours before cracking.

Residual deformation accumulation processes in thiokol sealants show up after contraction, especially at temperatures above 50°C. Thiokol sealants lose their shape, if subjected to a continuous stress. The more stress and temperature are, the more intensive creeping is observed. The increase of creeping is particularly evident at temperatures above 70°C and a pressure above 10 MPa. U30 M-grade sealant is the product, least subjected to creeping [1].

P-4-grade mixing solvent proved to suit best for regulation of viscosity during application of filled thiokol sealants, U30 M-grade sealant was taken for testing. Coatings have optimum properties at P-4 content of ~30%, taking cross-link degree and increase of diffusion characteristics into consideration. Therefore, it is possible to apply such compositions by airless spraying [152].

Thiokol sealants can be destructed by simple technology and form compositions with a distinct viscosity, if waste treatment is required. 2-mercaptobenzothiazol (captax) and dibenzothiazolyldisulfide (altax) are proposed for destructing agents [155]. Destruction mechanism in the presence of captax is supposed to be main chain decomposition via acetal bond opening (Eq.(3)):

$$(3)$$

Authors consider altax to participate in exchange reactions with thiokol's main chain. The reaction type is sulfide-disulfide exchange and the mechanism ii radical (Eq. (4)):

$$(4)$$

There are selected curing agents for formed regenerates of sealants, made of liquid thiokol and TPM-2 polymer. Such compositions have been suggested to be used as glues, sealants for threaded connections, as well as for mechanical engineering industry [155].

Thiokol sealants are also used for sealing at oil storages and gas stations [2, 3].

Liquid thiokols are used for plasticization and modification of epoxy Diane resin compositions [1, 2, 156]. Modified liquid thiokols with epoxy end groups have become more and more popular for these purposes in recent years. At first, there were ELP-3 and ELP-612 grades (Morton) and there are EPS-type aliphatic thioplasts (Akzo Nobel) at present thioplast EPS 15, EPS 25, EPS 70 and EPS 350 aliphatic thioplasts (Table 2.7) [157].

**TABLE 2.7**   Grades and Properties of Liquid Thiokols With Epoxy End Groups

| Properties | EPS-15 | EPS-25 | EPS-70 | EPS-350 |
|---|---|---|---|---|
| Viscosity, Pac(20°C) | 9–15 | 2–3 | 5–10 | 30–40 |
| Degree of branching, molar % | 2 | 2 | 0 | 0 |
| Structure | aliphatic | aliphatic | aliphatic | aliphatic |
| Oxygen content in epoxy groups, mass % | 1–1.5 | 2.1–2.9 | 4.6–5.0 | 4.5–5.0 |
| Epoxy equivalent, g/equivalent. | 1100–1500 | 500–600 | 280–350 | 300–370 |

Such oligomers can used both individually and in compositions with epoxy resins for making of coverings, paints, glues and sealants with fast curing rate at normal temperatures, adhesion to many surfaces, controlled flexibility, high impact viscosity, chemical resistance to many acids, bases and solvents and licensed for indoor applications as they do not smell as standard liquid thiokol sealants.

Small quantities of liquid thiokols are used as effective reactive plasticizers for butadiene-nitrile rubbers [158, 159]. They demonstrate plasticizing effect during mixing and, therefore, increase process ability of rubber compositions, oil-and-gasoline resistance, antifriction properties and frost resistance of related products. One must note that there were attempts to use liquid thiokols for production of tires. It were 1970s, when it was determined, that addition of 4–10 mass parts of liquid thiokol into rubber compositions of natural or isoprene rubber and butadiene-styrene rubber reduces reversion and heat build-up, increases dynamic properties and eliminates discoloration, that is an intrinsic quality of sulfur-containing compositions [160, 161]. Publications [158, 162–164] describe the mechanism of unsaturated rubber vulcanization by a liquid thiokol as well as the solid "DA"-grade thiokol and the influence of a molecular weight, functionality and the method of making on its effectiveness for vulcanization [158, 162–164]. There are also studies of how sulfuric vulcanizing group components and combinations of liquid thiokol with cobalt compounds influence the effectiveness of application of liquid thiokol for production of tire rubbers [165, 166]. Although there is no up to date information about industrial application of liquid thiokol as a vulcanizing agent, their prospective application attracts attention today [3, 162].

PSO sealants are mainly used in construction at the present time [3, 167, 168]. There are three general trends of application of sealing compositions in construction:
- production seams for sealing in large-panel construction;
- sealing of glass packets;
- road construction: sealing of temperature, deformation and shrinkage seams in concrete or asphalt pavements, landing strips and other constructions.

There are many types of sealants at the present time, following a broad range of requirements and application conditions and being suitable for substrates with various properties.

PSO sealants can cure and preserve good properties in a broad range of deviations from optimal curing agent dosages, as well as keep sealant components highly stable at storage before use. PSO sealants are highly sensible to a curing rate, depending on air humidity, on the one hand, and are fully independent of sealant's final properties on this factor on the other hand.

Above-listed disadvantages of PSO sealants make it possible to apply them in construction, where it's not always possible to make a precise weighting of components, adhere to temperature regimes and humidity during making and application of sealants.

Considering above data, the most preferable sealants for sealing inter-panel joints in building construction are thiol-containing polyesters (LT-1, SG-1, "Sazilast," "Thixoprol"). Such sealants are usually inclined to higher deformation, than thiokol-containing ones and suit for year-round application from –20 to +40°C.

TPM-2 sealants are of a low-modular type in their physico-mechanical properties. They have low thixotropicity. Their breaking strength is small, its value is within 0.15–0.5 MPa at relative elongation of 150–350% (in seams). They are highly adhesive to various surfaces (concrete, glass, steel, etc.) and are highly resistant to UV-radiation and ozone, highly thermally stable and can be used in a temperature range of –50 to + 80°C. The service life of sealants is at least 15 years. Major producers of such putties and properties of these products are given in Table 2.8. The important advantage of this sealant over thiokol ones is its process ability and workability at negative temperatures. Service properties of such putties are equal to those of construction materials, based on the AM–05 liquid thiokol, while they excel this thiokol in adhesive and elastic properties. However, TPM-2-based putties are highly tailored and designed for sealing interpanel seams of buildings, while their resistance at a constant contact with water is insufficient. That's why liquid thiokol putties also enjoy a firm demand on a construction market, as they are effective at a constant contact with water, for example, in horizontal adjoining Thiokol sealants excel TPM 2 polymer compositions in strength (the minimal conventional strength value of products from various producers is 0.2 MPa according to a technical documentation). However, their warranty storage period before use does not usually exceed three months. Comparing urethane sealants used for same purposes, PSO sealants can be applied on a humid surface without deterioration a sealing quality. Application of urethane sealants on a humid surface results in formation of a foamed layer on the concrete-sealant interphase boundary due to the reaction of isocyanate group with water ( $\sim2R\text{-}NCO + H_2O \sim R\text{-}NHC(O)NHR\sim + CO_2$ ). It leads to deterioration of waterproof properties.

Glass packets are extensively used both in civil and housing construction in the last 15–20 years. Their application increases air-tightness

**TABLE 2.8**  Producers, Grades and Properties of Sealants for Interpanel Seams

| Producer | Grade | Polymer | Viability, hours | Strength, MPa | Relative elongation, % | Deformability, % | Deposition temperature, °C | Color | Number of components |
|---|---|---|---|---|---|---|---|---|---|
| CJSC "SAZI", Russia | Sazilast 10 | Thiokol | >2 | >0,15 | >300 | 50 | -15 ÷ +35 | White | 1 |
| | Sazilast 201 | Thiokol | 3-12 | >0,4 | >200 | 50 | -15 ÷ +40 | Light gray | 2 |
| | Sazilast 202 | Thiokol | 3-12 | >0,4 | >200 | 50 | -20 ÷ +40 | Light gray | 2 |
| | Sazilast 21 | Thiokol | 3-14 | >0,2 | >150 | 25 | -15 ÷ +40 | Light gray | 2 |
| | Sazilast 22 | Thiokol | 3-12 | >0,2 | >150 | 25 | -20 ÷ +40 | Light gray | 2 |
| | Sazilast 23 | TPM-2 | 2-24 | >0,15 | >150 | 20 | -20 ÷ +40 | белый | 2 |
| | Sazilast 24 | Urethane | 2-24 | >0,2 | >150 | 30 | -20 ÷ +40 | белый | 2 |
| | Sazilast 25 | Thiokol | 2-24 | >0,25 | >200 | 50 | -15 ÷ +40 | белый | 2 |
| "KZSC" Co Ltd, Russia | SG-1K | TPM-2 | 2-24 | >0,15 | >150 | 50 | -20 ÷ +40 | Light gray | 2 |
| | LT-1K | TPM-2 | 2-24 | >0,15 | >150 | 50 | -20 ÷ +40 | White | 2 |
| | AM-05K | Thiokol | 2-24 | >0,2 | >150 | 25 | -5 ÷ +40 | Light gray | 2 |
| "Tenax" Co Ltd, Latvia | Elur T | Epoxyurethane | 5-7 | 0,35-0,45 | 400-450 | 50 | -15 ÷ +40 | White | 2 |
| | Oxyplast | Urethane | 2-7 | 0,35-0,45 | 200-250 | 25 | -10 ÷ +25 | Gray | 2 |
| | Thioplast | Thiokol | >2 | 0,35-0,45 | 150-180 | 25 | -10 ÷ +40 | Gray | 2 |
| CJSC "TSK" Russia | Elur-2 | Epoxyurethane | 3-8 | >0,2 | >150 | 25 | -10 ÷ +40 | White | 2 |
| | Tektor 202 | Urethane | 2-10 | >0,2 | >150 | - | -10 ÷ +40 | White | 2 |
| "Tremco", USA | Dymonic | Urethane | - | 0,4 | 275* | 25 | +5 ÷ +35 | White | 1 |
| | Dymonic NT | Urethane | - | 0,4 | 592* | 50 | +5 ÷ +35 | White | 1 |
| "Crasco", Russia | CII-grade sealant | Urethane | 24 | >0,15 | >150 | 25 | - | - | 2 |
| "Germoplast" Co Ltd, Russia | Unigex | Urethane | 3-8 | >0,2 | >300 | - | -25 ÷ +50 | Gray | 2 |

* – on blade samples

leads to a considerable decrease of heat loss and noise and increase of comfort as the natural illumination increases, moisture condensation on glass reduces and dust does not come inside. An external deformation seam is sealed to provide a reliable glass packet construction. Gas tightness is the main functional characteristic of a sealant, which determines its suitability for sealing of a glass packet. Such sealants are also supposed to be fast cured, have good deformation properties, long-term adhesion to glass and duralumin in conditions of alternation deformations and the effect of moisture in the temperature range from –50 to + 80°C. The viscosity of compositions should allow their machine-assisted application. Their properties, including technological characteristics, should be highly similar from batch to batch.

Thiokol sealants are used mainly abroad, as they have the lowest gas permeability among curable oligomers and considering above requirements. The leading producers of sealants for glass packets in the European market are Kommerling and Teroson (Germany) and CJSC "SAZI" in the Russian market.

According to information from O.K.N.A. Marketing Company, Russian producers took 12% of thiokol sealant market in 2005, while 64% of sealants, used for secondary sealing of glass packet, were thiokol-based [169] (see Figs. 2.3 and 2.4).

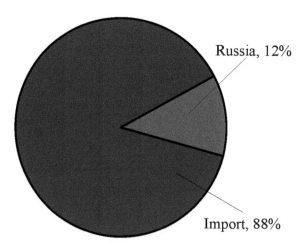

**FIGURE 2.3**    The share of foreign and Russian sealants.

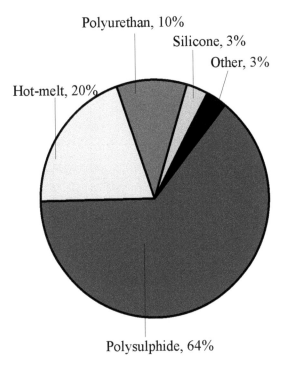

**FIGURE 2.4**   The share of sealants for secondary sealing of glass packets in Russia in 2005.

Properties and grades of sealants for glass packets are given in Table 2.9. A necessary prerequisite for good operating properties is a cohesion type of sealant's rupture, when its adhesion to glass or duralumin is estimated.

The requirements for thiokol sealants, designed for glass packets, are more severe, than for seam sealing compositions. These requirements consider the quality and stability of used raw materials, dispersity of fillers, manganese dioxide activity and volatility of plasticizer's vapors. First of all, these demands reflect mechanized production technology as well as required properties: adhesion to glass and duralumin, diffusion rate of stem through sealant's body and the rate of curing. Table 2.9 represents properties of thiokol sealants for glass packets, produced by some European companies and Russian producers.

**TABLE 2.9** Producers, Grades and Properties of Sealants for Glass Packets

| Parameters | Tenaglass-2 Tenax, Latvis | Thiover Fenzi, Italy | STIZ-30 JSC "SAZI", Russia | Terostat-998R Henkel TerOstOn, Germany | GD-116 Koemmerling, Germany | Totalseal 3185, LGF, France | Elastosil IG 25 Wacker Silicones, Germany | IS442, Tremco, USA |
|---|---|---|---|---|---|---|---|---|
| Polymer | Thiokol | Thiokol | Thiokol | Thiokol | Thiokol | Urethan | Silicon | Urethane |
| Paste density, g/sm³: - sealing paste - curing paste | 1,77 1,52 | 1,77 1,69 | 1,6 (mixed) | 1,71 1,6 | 1,78 1,51 | 1,2 (mixed) | 1,39 1,09 | 1,63 1,16 |
| Viability, min | 40-90 | 120-300 | 30-90 | 60+-10 | 60 | 30-40 | 60 | 30-60 |
| Minimal conventional strength at rupture, MPa | 1,0 | 0,9 | 1,0 | 1,37 | 1,65 | - | 0,9 | >0,8 |
| Minimal relative elongation, % | 50 | 50 | - | 145 | 125 | 200 | 100 | - |
| Adhesion to glass and duralumin, MPa* | - | - | 0,85-aluminum 1,05- glass | 1,0 | >1,3 | | | |
| Shore hardness A in 24 hours, conv. units | - | 50 | 50+4 | 45 | 42 | 40 | 42 | 45,15 |
| The ratio of component masses | 100 : 8,5 | 100 : 9,5 | 100:10 | 100:10 | 100:8,5 | 100:6,5 | 100:10 | 100:7 |

*A - cohesion type of destruction

Thiokol sealants are known to be used for sealing and repair of roof coverings (Sazilast 51, Sazilast 52, Sazilast 53), as well as for sealing concrete tanks at various waste disposal plants and waste water treatment infrastructure, thanks to their resistance to diluted acids and bases [170].

## KEYWORDS

- **composition**
- **oligomer**
- **polysulfide**
- **properties**
- **sealants**
- **vulcanization**

## REFERENCES

1. Averko-Antonovich, L. A., Kirpichnikov, P. A., & Smyslova, R. A. (1983). Polysulfide oligomers and related sealants (in Russian). Leningrad: Himija, 128 p.
2. Lucke, H. (1994). Aliphatic Polysulfides. Monograph of an elastomer Publisher Huthig and Wepf Basel, Heidelberg, New York. 191 p.
3. Li, T. S. P. (1995). Kauchuk i rezina (in Russian) 2, 9–13.
4. Smyslova, R. A. (1974). Liquid Thiokol Sealants (in Russian) M.: CNIITJeneftehim, 83 p.
5. Averko-Antonovich, L. A., Zareckij, Ja. S., & Kirpichnikov, P. A. et al. (1968). Kauchuk i rezina (in Russian) 4, 11–13.
6. Muhutdinova, T. Z., Averko-Antonovich, L. A., Kauchuk i rezina (in Russian) (1971). 12, 10–13.
7. Minkin, V. S., Averko-Antonovich, L. A., & Nefed'ev, E. S., et al. (1978). Vysoko-molek. soed (in Russian). Series B 20(6), 407–409.
8. Labutin, L. P. (1982). Rust Preventive and Sealing Materials Based on Synthetic Rubbers (in Russian). Leningrad: Himija 213 p.
9. Averko-Antonovich, L. A., Kirpichnikov, P. A., &Zareckij, Ja. S. (1965) Kauchuk i rezina (in Russian) 9, 20–23.
10. Barber, J. W., &Hahnele, P. J. (1986). Ind. and End. Chem. Prod. Res. and Dev, 25(2) P. 328–332.
11. Smyslova, R. A. (1984). Liquid Thiokol Sealants (in Russian) Moscow: CNIIT Jeneft-ehim. P. 67.
12. Synthetic Rubber (in Russian), (1983). Ed. by. Garmonov, I. V., Leningrad, Himija, 560 p.

13. Ramaswamy, R., &Achary, P. E. (1985). J. Appl. Polym. Sci. 30(9), 3569–3578.
14. Fuchs, W. (1987). Maler-und Lackirerhandverk 1, S.12–17
15. Smyslova, R. A., Shvec, V. M., & Sarishvili, I. G. (1991). Application of Curable Sealants in constructions Review of VNIINTI and on the economy of commercial construction materials (in Russian) Series 6 (2), 30 p.
16. Minkin, V. S. (1975). Dissertation of Candidate of Chemical Sciences (in Russian). Kazan: KSTU. 153 p.
17. Averko-Antonovich, L. A., Muhutdinova, T. Z., Minkin, V. S., & Jastrebov, V. N. (1974). Vysokomolek. Soed (in Russian).16A(8), 1709–1713.
18. Shljahter, F. A., &Novosjolok, F. B. (1983). Polysulfide Rubbers, in "Synthetic Rubber" (in Russian). Ed. by Garmonov, I. V., Leningrad: Himija. 470–488.
19. Sigjeru, O. (1975). The Chemistry of Organic Sulfur Compounds (in Russian), Moscow, Himija, 512 p.
20. Fettes, E. M., &Iorczak, J. S. (1950). Ind. and Eng. Chem, 22, 2217–2221.
21. Iorczak, J. S., &Fettes, E. M. (1951). Ind. and Eng. Chem, 43, 324–328.
22. Nasonova, P., Shljahter, R. A., Novoselok, F. B., & Zevakin, I. E. (1973). Synthesis and Physical Chemistry of Polymers (in Russian), Kiev, 11, 60–63.
23. Tavrin, A. E. (1969). Chemical sciences candidate's dissertation abstract. 02.00.06. Kazan. 7 p.
24. Tavrin, A. E., &Tejtel'baum, B. Ja. (1969). DAN SSSR (in Russian) 186(5), 1065–1068.
25. Beregovskaja, M. G., Nasonova, A., & Muljukova, S. G. (1960). Kauchuk i rezina (in Russian), 5, 37–39.
26. Henbest, H., Thomas, A. (1976). Chem. and Industry, 40 1097–1099.
27. Minkin, V. S., Averko-Antonovich, L. A., Jastrebov, V. N., &Muhutdinova, T. Z. (1987). Vysokomol. Soed. (in Russian), 16A(8), 1709–1713.
28. Nefed'ev, E. S., Minkin, V. S., Averko-Antonovich, L. A., & Kirpichnikov, P. A. (1981). Vysokomol. Soed (in Russian), Series. B, 23(8), 593–596.
29. Minkin, V. S., Nefed'ev, E. S., Averko-Antonovich, L. A., & Kirpichnikov, P. A. (1983). Izv. VUZov SSSR, Chemistry & Chemical Technology series (in Russian), Vol.26, 3, 348–351.
30. Korotneva, L. A., Belonovskaja, G. P., & Dolgoplosk, B.A. (1968). Vysokomol. Soed. (in Russian), Series B, Vol.10,1, 4–5.
31. Coates, B. C. G., ilbert, T. C. P., &Lee, I. (1992). Chem. Soc. Perkin Trans. 2, 1387.
32. Patent 4104189 USA, MKI C 08 F 28/00.
33. Patent Application, FRG 2557012, published 06.30.77. RZhH (in Russian) (1978). 9T 32 P.
34. Averko-Antonovich, L. A., Minkin, V. S., Nefed'ev, E. S., & Rubanov, V. E. (1978). Kauchuk i rezina (in Russian)10. P. 12–15.
35. Minkin, V. S., Averko-Antonovich, L. A., Safina, N. P. et al. (1984). Kauchuk i rezina (in Russian). 11, 9–11.
36. Brun, Ch., Buhrer, H. (1982). "Farbe und Lack" 88(2) 636–641
37. Suhanov, P. P. (1985). Chemical sciences candidate's dissertation Kazan, 202 p.
38. Averko-Antonovich, L. A., Kirpichnikov, P. A., & Prohorov, F. S. (1968). Kauchuk i rezina (in Russian), 5, 18–20

39. Minkin, V. S., Averko-Antonovich, L. A., & Kirpichnikov, P. A. (1975). Vysokomolek. Soed (in Russian) Series B, Vol. 17(1), 26–29.
40. Averko-Antonovich, L. A., Rubanov, V. E., & Klimova, L. I. (1977). Vysokomolek. Soed. (in Russian). Series A Vol. 19(7), 1593–1598.
41. Averko-Antonovich, L. A. (1978). Izv. VUZov, "Chemistry & Chemical Technology" series (in Russian). 21(2), 263–267.
42. Averko-Antonovich, L. A., Muhutdinova, T. Z. (1976). Izv. VUZov. "Chemistry & Chemical Technology" series (in Russian). 1, 164.
43. Averko-Antonovich, L. A., Kirpichnikov, P. A. (1978). Abstracts of International Conference on Caoutchouc and Rubber (in Russian), Kiev, 44–51.
44. Muhutdinova, T. Z., Averko-Antonovich, L. A., & Kirpichnikov, P. A. (1973). KSTU Transactions (in Russian), 52, 88–97
45. Shljahter, R. A., Apuhtina, N. P., &Nasonova, G. P. (1963). Dokl AN SSSR (in Russian), 149(2), 345–347.
46. Minoura, Y. (1955). J, Soc.Rubber. Jnd, Japan, 8, 399–407.
47. Ghatge, N. D. (1979) Rubber chem. and Technol. V. 57(4), R.744–754
48. Patent 3637574 USA, MKI C 08 G 25/00.
49. Patent 2000752 FRG, MKI G 02 B 21/36.
50. Patent of Japan 52–22967. (1977).
51. Author's Certificate 1306091, S 08 L 81/04.
52. Patent 4165426 USA, MKI S 08 G 75/04.
53. Patent 3923754 USA, MKI S 08 G 75/00.
54. Patent 4314920 USA, MKI S 084 93/00, MKI 260/24, RZhH (1982), 23T652P.
55. Patent 4263078 USA, (1981).
56. Patent 4311492 USA, (1982).
57. Patent 4165425 USA, (1979), RZhH 1980, 3T616P.
58. Patent Application, Japan 58–134123 RZhH (1984), 20T2023P.
59. Patent Application 60–215055, Japan, MKI S 08 L81/04.
60. Hiroyoshi Kuramoto. (1983). Toray Thiokol Company, Ltd., International Technican Meeting, Hakone, Japan.
61. Rahmatullina, G. M. (1980). Chemical sciences candidate's dissertation (in Russian): 02.00.06. Kazan. 182 p.
62. Kirpichnikov, P. A. (1979). Vysokomolek. Soed (in Russian) Vol. 21A.11. P. 2457–2468.
63. Elchueva, A. D. (1989). Chemical sciences candidate's dissertation (in Russian): 02.00.06. Kazan. 154 p.
64. Polikarpov, A. P., Shahbazov, G. M., Andreeva, G. I., & Averko-Antonovich, L. A. (1987). Synthesis and Properties of Polyesterurethane Elstomers (in Russian): VNI-ISK Transactions. Moscow: CNIITJeneftehim. P. 128–134.
65. Patent 1286281 France, MKI C 08 G 75/14.
66. Patent 975907 England, MKI C 08 G 75/04.
67. Averko-Antonovich, L. A., Kirpichnikov, P. A., & Romanova, G. V. (1969). KSTU Transactions (in Russian) Issue 40 Part 2. P. 54–62.
68. Minkin, V. S., Romanova, G. V., Averko-Antonovich, L. A., Skvorcova, O. V., & Kirpichnikov, P. A. (1975). Vysokomolek. Soed. (in Russian). Series B. Vol.17(11), 831–834.

69. Minkin, V. S., Romanova, G. V., &Averko-Antonovich, L. A., Vysokomolek. Soed (in Russian) (1975). Series A. Vol. 17(5), 1009–1113.
70. Minkin, V. S., Averko-Antonovich, L. A., Romanova, G. V., & Skvorcova, O. V. (1974) Transactions of Kazan Chemical & Techological Institute (in Russian). Issue: 54, 73–78.
71. Minkin, V. S., Averko-Antonovich, L. A., Romanova, G. V., & Vorotnikova, G. P. (1974) Transactions of Kazan Chemical & Technological Institute (in Russian). Issue 54, 68–72.
72. Author's Certificate 594027, MKI S08 L 81/04, "Sealing composition" (in Russian).
73. Arkina, S. N., Berlin, A. A., & Kuz'minskij, A. S. (1971). Transactions of International Conference of Caoutchouk & Rubber (in Russian). Moscow: Himija. 241–247.
74. Minkin, V. S., Suhanov, P. P., Averko-Antonovich L. A., & Kirpichnikov, P. A. (1989). Abstracts of All-Russian Conference on Radical Polymerization (in Russian) Gorkij, 168
75. Smyslova, R. A., Kotljarova, S. V. (1976). Handbook on Rubber Sealing Materials Moscow: Himija. 72 p.
76. Chernjak, K. I. (1970). Nonmetallic Materials in Marine Electrical and Radio Equipment (in Russian) Leningrad: Sudostroenie. 560 p.
77. Chernin, I. Z. (1971). Vysokomolek. Soed. (in Russian). Vol.13 B, 7, 502
78. Kardashov, D. A. (1973). Epoxy glue (in Russian). Moscow: Himija. 191 p.
79. Apuhtina, N. P. (1961). Polysulfide Polymers (Thiokols) Special rubbers (in Russian). Book of reviews, Moscow: 85–120.
80. Patent 1316600, France, MKI S 09 K 3/10.
81. Pavlov, G. N., &Jerlih, I. M. (1970). Kauchuk i rezina (in Russian) .11, 20–22.
82. Muhutdinova, T. Z., Shahmaeva, A. K., Gabdrahmanova, F. G., & Sattarova, V. M. (1980). Kauchuk i rezina (in Russian) 1 P. 12–15.
83. Polikarpov, A. P., Averko-Antonovich, L. A., Romanova, G. V., & Krasil'nikova, A. P. (1982), Kauchuk i rezina (in Russian).2, 27–29.
84. Polikarpov, A. P., Averko-Antonovich, L. A. et al., & Izvestija VUZov. (1982). Chemistry & Chemical Technology series (in Russian) Vol. 25(11), 1388–1392.
85. Polikarpov, A. P., Shahbazov, G. M., Andreeva, G. I., & Averko-Antonovich, L. A. (1987). Transactions of Institutes (in Russian). Kazan, KHTI, 128–131
86. Petrov, O. V., Nefed'ev, E. S., & Kadirov, M. K. (1998). Kauchuk i rezina (in Russian) 5, 13–19
87. Polikarpov, A. P. (1982). Chemical sciences candidate's dissertation (in Russian). 02.00.06. Kazan. 171 p.
88. Nefed'ev, E. S. (1991). Chemical sciences doctor's dissertation (in Russian) Kazan', 233 p.
89. Ashihmina, L. I., Jamalieva, L. N., Nefed'ev, E. S., & Averko-Antonovich, L. A. (1988). Kauchuk i rezina (in Russian)12, 19–21
90. Ashihmina, L. I. (1989). Technical sciences candidate's dissertation (in Russian). Kazan' 168 p.
91. Nefed'ev, E. S., Ashihmina, L. I., Ismaev, I. Je., Kadirov, M. K., Averko-Antonovich, L. A., & Il'jasov, A. V. (1989). Dokl. AN SSSR (in Russian) 304(5), 1181–1184
92. Kirpichnikov, P. A., Il'jasov, A. V., Kadirov, M. K. et al. (1986). Izv AN SSSR. Chemical series (in Russian), 12, 2824.

93. Kirpichnikov, P. A., Il'jasova, A. V., Kadirov, M. K. et al. (1987). Dokl. AN SSSR (in Russian) 294(4), 910–913.
94. Patent 5663219 USA, MKI S08 K3/20.
95. Elchueva, A. D., Tabachkov, A. A., Ioffe, D. S., & Liakumovich, A. G. (2001). Kauchuk i rezina (in Russian), 1, 15–16.
96. Patent 3446780 USA, MKI S 08 G 12/20.
97. Cherkasova, L. A., Sotnikova, Je. N., Shvecova, E. I. et al. (1976). Synthesis and Physical Chemistry of Polymers (in Russian) Kiev: Naukova Dumka. 9, 72–76.
98. Shitov, V. S., Matveev, G. V. (1980). Elastomer Polyurethane sealants (in Russian) M: CNIIT Jeneftehim. 64 p.
99. Rahmatullina, G. M., Averko-Antonovich, L. A., & Kirpichnikov, P. A. (1979). Catalysts for reactions of oligothiols with oligoisocyanates (in Russian) Kazan: KHTI. deposited in ONIITJehim.2929179, 12 p.
100. Homko, E. V., Myshljakovskij, L. N., & Tonelli, K., Paint-and-Lacquer Materials (in Russian), 2884(5), 9.
101. Hastings, G. W., Johnston, D. (1971). Brit. Polym. J. 3(2), 83–85
102. Patent 3440273 USA, MKI C 08 G 12/20.
103. Patent Application 3407031 Germany, MKI S 08 G 18/10, S 08 G 18/12.
104. Patent Application, Japan 63–145319 MKI S 08 G 18/50, RZhH (1989), 20T196P.
105. Sharifullin, A. L. (1991). Chemical sciences candidate's dissertation (in Russian): 02.00.06. – Kazan. 154 p.
106. Apuhtina, N. P., Novoselok, F. B., Kurovskaja, L. S., & Ternavskaja, G. K. (1970). Synthesis and Physical Chemistry of Polymers. Kiev: Naukova Dumka. 6, 141–143.
107. Minkin, V. S., Averko-Antonovich, L. A., Kirpichnikov, P. A., & Suhanov, P. P. (1989). Vysokomol. Soed. (in Russian), 31(2), 238–251.
108. Minkin, V. S., Averko-Antonovich, L. A., Romanova, G. V., & Kirpichnikov, P. A. (1975) Vysokomol. Soed. (in Russian). 17 B(5), 394–396.
109. Polikarpov, A. P., Romanova, G. V., Averko-Antonovich, L. A., & Smyslova, R. A. (1984). Chemistry and Technology of Elementorganic Polymer Compounds, Kazan, KHTI, 61–64.
110. Saunders, J. H., Frish, K. K. (1968). Polyurethane Chemistry (Russian Translation), Moscow Himija, 470 p.
111. Patent Application 3707350 FRG, MKI C 08 L 75/04, C 08 L 81/00, Teroson GmbH, RZhH (1989), 10T376P.
112. Selivanov, A. V. (1985). Chemical sciences candidate's dissertation (in Russian): 02.00.06. Kazan. 175 p.
113. Patent 5319057 USA, MKI C08 G18/18. (1996).
114. Novoselok, F. B., Rowina, N. A., & Apuhtina, N. P. (1976). Synthesis & Properties of Urethane Elastomers (in Russian), Leningrad, Himija, 23–25.
115. Patent 5679756 USA, MPK C08 G18/10. (1999).
116. EPV (ER), Application 0281905, MKI 4 C 08 L 75/08, C 08 G 18/18, C 09 D 3/72.
117. Patent 4797463 USA, MKI C 08 G 18/10, Teroson GmbH.
118. Strengthening of Elastomers, ed. by Krausa, J. (in Russian), Moscow, Himija, (1968). 463 p.
119. Hristova, D., Mladenov, I.v.,Edreva, M.,& Him, et al. (in Russian), (1988). 6 254–256.

120. Muhutdinova, T. Z., Averko-Antonovich, L. A., Kirpichnikov, P. A., & Muhametsali-hova, V. G. (1972). KSTU Transactions 50, 153–160.
121. Minkin, V. S., Averko-Antonovich, L. A., & Bezrukov, A. V. (1975). Izv. VUZov Chemistry and Chemical Technology series (in Russian) 18(12), 1953–1956.
122. Averko-Antonovich, L. A., Muhutdinov, A. A., Muhutdinova, T. Z., & Hismatullin, R. A. (1977). Izv. VUZov (in Russian) 20(4), 564–567
123. Petrov, O. V. (1999). Chemical sciences candidate's dissertation (in Russian), Kazan 106 p.
124. Nistratov, A. V. (2006). Technical sciences candidate's dissertation (in Russian), Volgograd, VGTU.
125. Ultracarb in polysulfide sealants Eur. Adhes and sealants, (1989). 3(6), 20.
126. Skelhorn, D., &Lee, R., Eur. Adhes. and sealants, (1988). 5(4), 30–32.
127. Fedjukin, N. D. (1986). Technical sciences candidate's dissertation (in Russian), Moscow, Moscow Institute of Fine Chemical Technology 134 p.
128. Novakov, I. A., Nistratov, A. V., Vaniev, M. A., & Luk'janichev, V. V. (2006). Klei, germetiki i tehnologii (in Russian), 2 P.15–18.
129. Novakov, I. A., Nistratov, A. V., Vaniev, M. A., Luk'janichev, V. V., Zerwikov, K., & Klei, Ju. (2006). germetiki i tehnologii (in Russian), 3, 23–28 .
130. Author's Certificate. NRB 19689, (1978). MKI S 08 G 75/04.
131. Nuralov, A. R., Breeva, G. I., & Gulimov, A. G. (1977) Stroitel'nye materialy (in Russian). 9, 8.
132. Mladenov, I. T., Markov, M. K., Todorova, D. D., & Todorov, S. N. (1984). Kauchuk i rezina (in Russian). 12. P. 8–11.
133. Averko-Antonovich, L. A., Muhutdinova, T. Z., & Kirpichnikov, P. A. (1975). i dr. Kauchuk i rezina (in Russian). 4. P. 18–20.
134. Ramaswamy, R., Sosidharan Achary P. (1982). Adhes.Joints: Format., Charact. and Test. Proc. Jnt. Symp., Kansas City, Mo., 12–17 Sept., New-York, London, 31–40
135. Rudenko, N. I., Baranovskaja, N. B., & Zel'bert, L. E. (1964). Kauchuk i rezina (in Russian). 2. – P.28–29.
136. Minkin, V. S., Deberdeev, R. Ja., Paljutin, F. M., Khakimullin, Yu. N. (2004). Commercial Polysulfide Oligomers: Synthesis, Vulcanization, Modification, Kazan, Novoe znanie, 175 p.
137. Patent 3653959 USA, MKI S 08 G 75/04 V.
138. Minkin, V. S., Averko-Antonovich, L. A., Romanova, V. G., & Kirpichnikov, P. A. (1975).Vysokomolek. Soed (in Russian) Vol. 17B.-5. – P. 394–396.
139. Minkin V.S., Romanova G.V., & Averko-Antonovich, L. A. et al. (1982). Vysokomolek. Soed (in Russian). 24B(7), 806–809.
140. Rubanov V.A. Technical sciences candidate's dissertation (in Russian), Kazan', KHTI, 1978.
141. Minkin, V. S. (1977). NMR in industrial polysulfide oligomers (in Russian) // Kazan, "ABAK", 222 p.
142. Patent 32974732, USA, RZhH (1968), 16S581P.
143. Patent 993963, Great Britain, RZhH (1966), 9S721P.
144. Patent 4070328, USA, MKI C 08 G 75/04.
145. Patent 1457872 Great Britain, MKI S 08 G.

146. Valeev, R. R. (2004). Technical sciences candidate's dissertation (in Russian), Kazan, KSTU.
147. Author's Certificate. 1054397, Open Images & Trademarks (in Russian), 1983, 42, 116.
148. Author's Certificate. 1069411, 1983,S09K 3/10,S08 81/04.
149. Author's Certificate. 1036039, 1983, S09K 3/10, C08K 5/49.
150. Patent Application 2623088, Germany, published in 11.24.77.
151. Mudrov, O. A., Savchenko, I. M., & Shitov, V. S. (1982). Handbook on Elastomer Coatings and Sealants Used in Shipbuilding (in Russian) Leningrad: Sudostroenie.184 p.
152. Novakov, I. A., Nistratov, A. V., Semenov, Ju. V., Vaniev, M. A. et al. (2005). Klei,germetiki, tehnologii (in Russian), 8, 17–20.
153. Henhela, P. J., Ind. and Eng. Chem. Prod Res and Dev, (1985). 25(2), 321–328.
154. Wischmann, K. B. (1982). Amer.Chem.Sos.Polym. Prepr, 23(1), 246–247.
155. Markov, V. V., Reznichenko, S. V., Evreinov, Ju. V., Ionov, Ju. A., & Morozov, Ju. L., (2005). Kauchuk i rezina (in Russian), 3, 24–28.
156. Rees, T. M., & Wilford, A. I. (1989). Oil and Colour Chem. Assoc., 72(2), 66–71.
157. "Aczo Nobel" company's datasheet.
158. Averko-Antonovich, L. A., & Averko-Antonovich, I. Ju. (1994). Sulfur Polymers & Copolymers as Modifiers and Vulcanizing Agents for Rubbers. A topical survey of synthetic rubber industry (in Russian), Moscow, CNIITJeneftehim, 73 p.
159. Ulitina, O. N. (1981). Tire production (in Russian), RTI & ATI, 3, 4–6.
160. Volfson, S. I., Levit, E. Z., & Averko-Antonovich, L. A. (1971). KSTU Transactions (in Russian), Vol. 46, 28–32.
161. Stevens, W. H. (1965). Rubber Plast Age, 36(12), 4696–4703.
162. Ajupov, M. I., Vol'fson, S. I., Mirakova, T. Ju., Nefed'ev, E. S., & Khakimullin, Yu. N. (2001). Some Aspects of Influence of Formulation Factors on the Strength Properties of Rubbers (in Russian), Kazan, KSTU, 89 p.
163. Hofman, R. F., &Schultheis, I. I. (1976). Elastomer, 11(7), 30.
164. Muhutdinov, A. A., Zelenova, V. N., Shabanova, A. D., & Fetisova, L. M. (1980). Kauchuk i rezina (in Russian), p.
165. Prokof'ev, Ja. A., Potapov, E. Je., Saharova, E. V., & Salych, G. G., (1999). Kauchuk i rezina(in Russian), 2, 20–23.
166. Prokof'ev, Ja. A., Potapov, E. Je., Saharova, E. V., Salych, G. G., & Kauchuk i rezina (in Russian), (1999). 3, 9–11.
167. Khakimullin Yu. N. (1977). NPK Transactions "Production & Consumption of sealants and Other Construction Compositions: Present State & Prospectives" (in Russian). Kazan, 27–39.
168. Khakimullin, Yu. N. (2003). Technical sciences doctors's dissertation (in Russian), Kazan, KSTU,.
169. http://www.oknamar.ru
170. CJSC "SAZI" datasheet, http.//www.sazi.ru.

# CHAPTER 3

# VULCANIZATION MECHANISMS FOR POLYSULFIDE OLIGOMERS AND THE INFLUENCE OF A VULCANIZING AGENT'S NATURE ON THE PROPERTIES OF SEALANTS

## CONTENTS

## 3.1   THE INFLUENCE OF MANGANESE DIOXIDE'S STRUCTURE ON VULCANIZATION OF LIQUID THIOKOLS

Manganese dioxide is known to oxidize alcohols by a radical mechanism [1–3]. Complexation processes exert rather important influence here. They involve not only $MnO_2$ ions but ions of other metals as well, which are always present in manganese dioxide as impurities.

Metal dioxides can interact with PSO forming mercaptide bonds. There is also another viewpoint on vulcanization mechanism. According to it, mercaptide bonds do not form in a cured system or their content is small, because formation of mercaptides requires either a metal atom withdrawal from a crystal lattice or C-S bond opening. It requires extra energy, which is absent in PSO vulcanization conditions [4, 5].

Attention is also attracted by the problem of determination of how the structure of industrial manganese dioxide influences the curing rate and properties of sealants, which are made of commercial liquid thiokols.

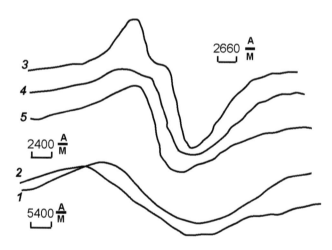

**FIGURE 3.1**   EPR spectra of initial (1), the vulcanizing paste number 9 (2) and liquid thiokol vulcanizates with a vulcanizing agent concentration n=1(3), 2(4) and 3(5) at 290°K.

Figure 3.1 demonstrates EPR spectra of the vulcanizing paste number 9, prepared using $MnO_2$ and liquid thiokol vulcanizates with various concentrations of this vulcanizing agent. The values of EPR spectra parameters made for $MnO_2$ and liquid thiokol vulcanizates are given in Table 3.1.

EPR spectra parameters show, that both $MnO_2$ and the related paste number 9 give EPR signal from $Mn^{4+}$ ions. The values of g-factor, width and form of this line are identical to observed for $MnO_2$ or $Mn^{4+}$ in glass [6].

The presence of a wide EPR line is caused by a strong covalent bond of $Mn^{4+}$ ion with oxygen atoms. $Mn^{4+}$ ion, is known to be in a $3d^3$ electron state [6], which is similar to that of $V^{2+}$ and $Cr^{3+}$ ions. One of the facts, confirming the presence of such ions, can be the decrease of g-factor value with the increase of temperature [7]. This effect is observed experimentally; when $MnO_2$ and the vulcanizing paste number 9 are used.

**TABLE 3.1**  EPR Spectra Parameters for $MnO_2$ and Liquid Thiokol, Vulcanized by $MnO_2$

| Sample | g-factor $\pm 0.001(290K)$ | g-factor $\pm 0.001(77K)$ | (290K) $d H \times 10^{-2}$, A/M | (77K) $dH \times 10^{-2}$, A/M |
|---|---|---|---|---|
| 1.  $MnO_2$ (powder) | 1.956 | 1.992 | 2954.9 | 357.4 |
| 2.  $MnO_2$ (past number 9 with 40% of the main substance) | 1.958 | 1.992 | 2460 | 358.4 |
| PSO vulcanized by $MnO_2$ (n-$MnO_2$ Is taken in excess comparing stoichiometric quality) | | | | |
| 3.  n=1.0 | 2.062 | 2.005 | 328.8 | 435.6 |
| 4.  n=2.0 | 1.961 | 1.988 | 332.75 | 397.6 |
| 5.  n=3.0 | 1.951 | 2.004 | 339.7 | 369.6 |

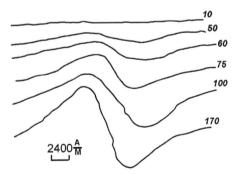

**FIGURE 3.2**   The kinetics of liquid thiokol vulcanization by manganese dioxide (n=1.5). Numbers on curves indicate reaction time in minutes.

Figure 3.2 represents EPR spectra, taken in radio spectrometer resonator during vulcanization of liquid thiokol by manganese dioxide. EPR signal appeared in 50 minutes after reaction initiation; its intensity was rapidly growing along with the process and became constant in four hours.

According to calculated parameters of EPR spectra (g=2.0111 ± 0.0013, line's width and form) the observed signal indicated the presence of manganese ions with the oxidation rate $Mn^{2+}$ [6, 7].

EPR spectra for PSO vulcanizates, prepared with various content of an oxidant at varied temperature, are demonstrated in Fig. 3.3, and their parameters are given in Table 3.1. Vulcanized samples show a complicated EPR line at a room temperature. This complication disappears, when $MnO_2$ content increases, as well as temperature is reduced down to 77°K. Observed values of EPR spectra parameters and line shapes of liquid thiokol vulcanizates indicate $Mn^{4+} ® Mn^{2+}$ transition.

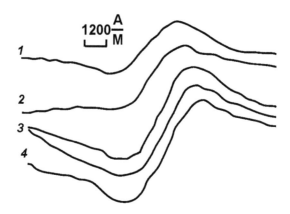

**FIGURE 3.3**   EPR spectra of initial $MnO_2$ (1) and liquid thiokol vulcanizate with the excess of a vulcanizing agent of n=1 (2), 2 (3), 4(4) at 77°K.

Therefore, there is an experimental confirmation of manganese ions transitions during oxidation of liquid thiokol by $MnO_2$. Obtained information provides a deeper review of oxidation mechanism and proposes promoting effect of water and other polar additives on PSO oxidation [8, 9].

A hyperfine structure for $Mn^{2+}$ ions in solid systems and powders is known to be observed only for samples with $N_p \sim 1.5 \times 10^{+17}$ $1/sm^3$ [10]. Such a concentration can take place if $Mn^{2+}$ ions are irregularly distributed inside a sample. Comparison of signal intensities of studied vulcanizate'

EPR spectra with those of a reference sample can show, that the concentration of paramagnetic centers $N_p$ in vulcanizate is ~ $5 \times 10^{18}$ $1/cm^3$. Therefore, a "blurred" fine structure is not observed in vulcanizate and line's shape is just slightly more complicated. The shape of EPR spectra for vulcanizate is also influenced by the ion distribution pattern in bulk polymer and localization places of ions.

Mn$^{2+}$ ions in vulcanizates seem to unite in ion-exchange pairs, making so-called "cluster" areas. This is similar to their distribution on powders, glasses, silica gels and resins. [11–13]. This fact is confirmed by reduction of EPR signal intensity with temperature [14], as well as by the change of a line shape, similar to observed in [13]. The fact that EPR line width little depends on manganese dioxide concentration (Table 3.1) can be explained by formation of microphases with adsorbed oxidizer on a surface of macromolecules, same as for Mn$^{2+}$ in silica gels [7]. EPR line width is in this case stipulated by magnetic interactions between ions, mainly belonging to those microphases, which slightly interact with each other. If it is true, than differences in MnO$_2$ dispersity might modify EPR line in PSO vulcanizate. Indeed, experimentally observed dH values are within 300e,450 e for MnO$_2$ oxidizedvulcanizate with various fractional compositions.

EPR spectra of vulcanizate and their sol-fractions are similar, but the last ones are less intensive, though EPR spectra in these systems are also not allowed for Mn$^{2+}$due to a sufficiently high system viscosity and increased concentration of paramagnetic centers (above the critical value). But the fact of Mn$^{2+}$ ions detection in a sol fraction confirms a possibility of localization of Mn$^{2+}$ ions on polymer chains. Therefore, Mn$^{2+}$ ions are irregularly distributed along a polymer surface and form microphases of adsorbed ions.

Adsorbed microphases of Mn$^{2+}$ ions in two neighboring planes can yet interact with each other via counter ions or water molecules [10, 14]. This interaction is especially strong with Cl$^-$ or OH$^-$ ions [10].

The Lorentz shape of EPR line, taken for Thiokol vulcanizate confirms the exchange style of interaction between microphases of adsorbed manganese ions, which are in different surfaces and substantially far from each other [10, 11]. Exchange processes in PSO vulcanizate can involve water molecules, and polymer moisture (0.2%) can contribute to interaction as well as water, forming during PSO vulcanization in the sulfhy dryl group oxidation reaction. Therefore, we can suppose, that as Mn$^{2+}$ ions accumulate in polymer, water molecules will favor oxidation of HS-groups due toarising interaction, in other words they will activate PSO vulcanization.

G-factor values of vulcanizates close to 2.0 may indicate, like in glasses [11, 14], manganese ions in networks ~S-Mn-S~ of mercaptide bonds. If manganese ions are in the network, thermal resistance of vulcanizate should drop. This fact is indeed observed when their thermal resistance is studied [1, 15].

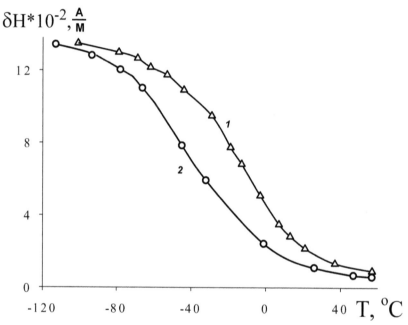

**FIGURE 3.4** Temperature dependencies of NMR line width for liquid Thiokol vulcanizate, cured by a powder-like manganese dioxide (1) and the paste number 9 (2).

Figure 3.4 demonstrates temperature dependencies of NMR line width for liquid thiokol vulcanizate, cured by the paste number 9 and powder-like manganese dioxide. The analysis of obtained dependencies shows, that the interval of fast NMR line narrowing is approximately similar for both vulcanizate, although the form of dH = f(t⁰) dependency is individual for these compounds. Fast NMR line narrowing indicates in both cases the start of segmental motion processes in vulcanizate [13].

One must note, that when manganese dioxide is used in the form of a vulcanizing paste, the temperature dependence of NMR line width shifts to a higher temperature area.

Temperatures of NMR line narrowing phenomena, determined by methods from [16, 17] ($T^N$), are −27°C for vulcanizates with a plasticizer and −45°C without a plasticizer correspondingly.

Such a difference in the behavior of vulcanizate can be explained by specificity of processes of network formation involving polysulfide oligomer at the presence of a crystalline manganese dioxide. Its introduction into a composition in a form of a homogeneous paste, which distributes throughout the reaction volume easily and uniformly, favors more complete conversion of oligomer end groups, than for the same quantity of a powder-like oxidizer. It affects the effective network density value, which differs ten times for different methods of oxidizer's introduction (Fig. 3.5). A better conversion of oligomer HS-groups in systems with a plasticizer favors increase of flexibility of macromolecules, while addition of a powder-like oxidizer reduces molecular mobility, due to the specific adsorption of macromolecules on solid $MnO_2$ surface.

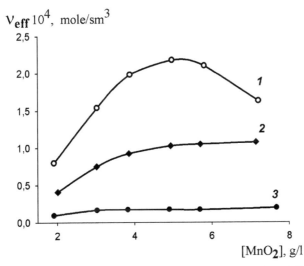

**FIGURE 3.5** The dependence of effective network chain density of liquid Thiokol vulcanizate on the content of the paste number 9 (1) and $MnO_2$ powder (3) ($MnO_2$, g/l of reaction volume); the dependence of chemical network chain density on the content of the paste number 9 (2).

Publications [3, 5] report about the role of adsorption phenomena for vulcanization of polysulfide oligomers by $MnO_2$. They propose to describe this process by chemical equations, considering adsorption-desorption phenomena.

Adsorptive interactions in systems with a powder-like oxidizer lead to a considerable heat generation during vulcanization. Thermal effects of vulcanization with a paste-like oxidizer are considerably lower and their corresponding values are (dH/s, J/g):

| Paste 9 | $MnO_2$ (powder) |
|---------|------------------|
| 1279,08 | 3852,44 |
| 2687,53 | 6927,62 |
| 2133,08 | 9346,74 |
| 2784,07 | 5560,70 |

Thermal effect of vulcanization is contributed by oxidation of end groups and adsorption-desorption effects. The last ones are especially important, when a powder-like cutting agent is used. It exerts a noticeable influence on heat generation during vulcanization, because a polymer network is not created in this case and formed vulcanizate swell in toluene so much, that it's not possible to determine equilibrium contraction modulus. Application of a paste-like manganese dioxide favors formation of network chemical chains, their concentration changes in symbiosis to $MnO_2$ content in a system (Fig. 3.5, curve 3).

The presence of a plasticizer in a system can, in turn, influences the number of donor-acceptor bonds, whichmay emerge in coordination of polymer chain atoms with manganese atoms. This fact is confirmed by differences in observed NMR line shapes for polysulfide oligomer vulcanizate, cured by a powder-like $MnO_2$ in the temperature range of $-12°C$ , $-60°C$. NMR line is single-component there, while it is supposed to be complicated for vulcanizate with a plasticizer.

Calculated values of second moments at lower temperatures ($-90°C$, $-100°C$) for vulcanizate, prepared using a powder-like manganese dioxide and a vulcanizing paste, proved to be $20.1 \times 10^{-4}$ $(A/m)^2$ and $26 \times 10^{-4}$ $(A/m)^2$ correspondingly. Second moment growth for the last formulation can be explained by the intensification of intermolecular interaction in vulcanizate, because of coordinative bonding, which has been previously reported in Ref. [15]. Activation energy values have been obtained from temperature dependencies of NMR line width. They indicate the start of a well-developed segmental motion in polymers, their values proved to be 24.8 kJ/mole and 27.2 kJ/mole for vulcanizate, prepared with a powder-like manganese dioxide and a vulcanizing paste correspondingly.

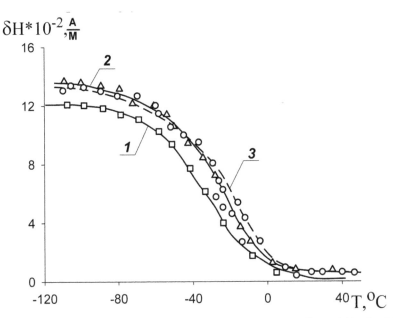

**FIGURE 3.6** Temperature dependencies of absorption NMR line width for PSO vulcanizate: 1- $MnO_2$ (n=1.0), 2- $MnO_2$ (n = 2.0).

It also possible to determine the mechanism of PSO vulcanization by manganese dioxide in the study of NMR line width temperature dependencies of PSO vulcanizates cured by various quantities of $MnO_2$.

Temperature dependencies of absorption NMR line width for PSO vulcanizate, cured by various quantities of $MnO_2$, are presented in Fig 3.6. All vulcanizate demonstrate rapid narrowing of absorption line width in the temperature range of −60.0°C. The variation of a vulcanizing agent quantity ($1 < n < 4$) exerts little influence on the temperature behavior of absorption line width. Figure 3.7 represents dependencies of molecular motion correlation times "lgtc* on 1/T," calculated out of temperature dependencies dH = $f$(T), according to methodology in Ref. [16]. Activation energy values for analyzed types of motion in vulcanizate are 26.9 kJ/mol for $n$ = 1.0 and 32.8 kJ/mol for $n$ = 2.0.

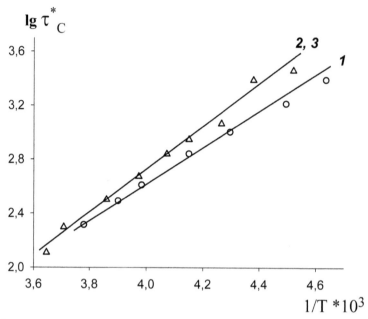

**FIGURE 3.7** The dependence of lgtc*–1/T for PSO vulcanizate: $1 - MnO_2$ (n=1.0), $2 - MnO_2$ (n = 2.0) and $Na_2 Cr_2 O_7$, (n = 2.0).

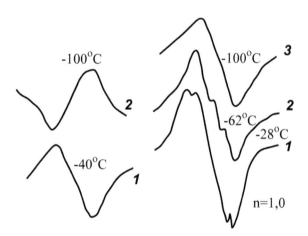

**FIGURE 3.8** NMR spectra of liquid thiokol vulcanizates, cured by MnO2 at various temperatures.

Figure 3.8 demonstrates shapes of NMR spectra, obtained for PSO vulcanizates. The most interesting fact is the spectra of vulcanizates cured by equimolar $MnO_2$ quantity ($n = 1,0$), which have the shape of single-component curves, while the excess of a vulcanizing agent creates two-component absorption lines. A complicated NMR line can be usually observed for manganese dioxide samples with uniform composition. Therefore, experimentally observed NMR spectra (Fig. 3.8) confirm the proposed mechanism of PSO oxidation by manganese dioxide. According to this mechanism, the surface of a crystalline vulcanizing agent contains adsorbed macromolecules, which are bound with oxidant not only by Vander Waals forces, but by weak mercaptide bonds as well. These bonds are likely to give a complicated shape of NMR absorption line. The more excess of $MnO_2$ system contains, the more adsorption-desorption mechanism contributes to vulcanization process.

One must note, that complicated NMR spectra for vulcanizates, cured by $MnO_2$, appearbelow the temperature of $-12$, $-14°C$ and exist up to $-62°C$. Further decrease of temperature leads to disappearing of narrow NMR spectral component and adsorption line becomes single-component again.

The study of how manganese oxide (IV) influences the curing rate of commercial polysulfide oligomers is of particular practical interest. PSO curing in practice often assumes various activity of constituent manganese dioxide [18]. PSO vulcanization leads to elongation of chains and their scanty cross-linking, when oxidizing mercaptan end groups of linear chains and their long-chain branching form during oligomer synthesis. It results in decrease of macromolecular mobility and related spin-spin relaxation time $T_2$. The latter parameter, as it has been shown in Refs. [19, 20], decreases down to a distinct minimal value and decreases in symbiosis with the network chain density of vulcanizates. It is a reliable structural and kinetic parameter for the analysis of vulcanization of liquid thiokols and related compositions [1].

Table 3.2 provides data on the composition and structural characteristics of commercial manganese dioxide batches.

**TABLE 3.2**   The Composition and Structural Characteristics of Commercial Manganese Dioxide Batches.

| Commercial batch | $MnO_2$ content, % | EPR line width dH, Gs (±0,2) (290K) | g-factor value (±0,001) | Line shape |
|---|---|---|---|---|
| 1 | 80 | 560 | 1.956 | Complicated |
| 2 | 78.2 | 1890 | 1.953 | Simple |
| 3 | 76.8 | Narrow compo- nent-620. wide com- ponent −2013.2 | 1.956 | Complicated |
| 4 | 82.7 | 1556.3 | 1.954 | Simple |

EPR spectra of commercial $MnO_2$ batches are similar to spectra, shown on Fig. 3.1. EPR spectra of manganese dioxide and related vulcanizing pastes are absorption signals, indicating the presence of $Mn^{4+}$ ions in glass or polymers [6, 7]. EPR line broadening is stipulated by the covalent bond of $Mn^{4+}$ ions with oxygen atoms, while its electron state is $(3d^3)$, that is similar to those of $Cr^{3+}$ and $V^{2+}$ [1, 6]. The maximal value of a dH resonance line is observed for samples 2 and 4 in Table 3.2. Samples 1 and 3 have the most explicitly complicated EPR line shape. It is significant, that spectral parameters of manganese dioxide and related vulcanizing pastes display same tendencies in the variation of a resonance line width. Meanwhile, vulcanizing pastes have considerably less wide resonance lines, because of a plasticizer (dibutylphthalate). The width and shape of EPR spectra resonance lines have been used as quantitative criteria of manganese dioxide activity, as these parameters correlate with the mobility and distribution of manganese ions ($Mn^{4+}$) in oxidant [21]. Figure 3.9 demonstrates kinetic curves of PSO vulcanization by manganese dioxides of various activities (curves 1 and 2).

Comparison of data in Table 3.2 and Fig. 3.9 shows, that observed differences in mobility and localization of ions of the base oxidant really influence vulcanization kinetics of commercial 1-grade Thiokol, when it is cured by $MnO_2$ (samples 1 and 2). The activity of sample 1 is almost twice higher, than the activity of sample 2. The effective vulcanization rate constant ($k_1 \times 10^3$ min$^{-1}$) value is 3, 7 for sample number 1 and 7, 8 for sample number 2 correspondingly. An oxidant appears to be active for industrial batches of Thiokol's with various molecular weights (Fig. 3.9, curve 3).

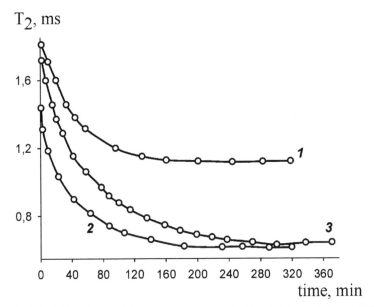

**FIGURE 3.9**  Vulcanization kinetic curves for industrial batches of liquid Thiokol. Manganese dioxide of various activity is used (curves 1 and 2, Thiokol of grade 1, curve 3, Thiokol of grade NVB-2).

There is strict correlation between EPR line width and the activity of commercial vulcanizing agents for all the studied batches of manganese dioxide. Vulcanizing agents with the line width of up to 600–700 Gs are relatively active in PSO vulcanization reaction. Oxidants with the line width above 1000 Gs are characterized by small values of kinetic parameters.

Active batches of $MnO_2$ give a complicated shape to observed EPR line, due to the presence of $Mn^{4+}$ ions with various mobility (Fig. 3.1). Indeed, a broad anisotropic $Mn^{4+}$ ion adsorption line is overlaid by the intensive narrow spectral component of the most mobile $Mn^{4+}$ ions. Their content in initial manganese dioxide can be 20–25% of total $Mn^{4+}$ content. That is why, despite broad EPR lines of commercial $MnO_2$ batches; their activity is relatively high because of more mobile $Mn^{4+}$ ions (the narrow spectral component) in a sufficiently high concentration, which are able to exert a crucial effect of PSO oxidation rate that is indeed observed for samples 1 and 3.

The analysis of obtained structural parameters makes it possible to estimate the activity of commercial manganese dioxide batches before PSO vulcanization. Differences in mobility, localization and concentration of mobile $Mn^{4+}$ ions in a vulcanizing agent (powder or paste-like) are likely to change PSO vulcanization rate. The latter one mainly determines the viability of thiokol compositions.

The influence of curing rate on properties of two types of sealants has been estimated to confirm above proposals: U-30 M (compositions 1 and 2) and AM-05 (compositions 3 and 4). These sealants differ in the nature of a filler (carbon black P-803 and chalk correspondingly) and the presence of E-40 epoxy Diane resin in AM-05 [22]. Compositions 1 and 3 were cured by manganese dioxide (sample 1, Table 3.2), compositions 2 and 4 were cured by manganese dioxide (sample 2, Table 3.2). Sealants were cured for 48 h at 70°C after viability loss.

Table 3.3 represents properties of thiokol sealants, prepared with varied curing rate.

**TABLE 3.3**    The Properties of Thiokol Sealants with Various Curing Rate

|   | Sealant grade | Conventional strength at rupture, MPa | Relative elongation, % | Adhesion to duralumin, kN/m | Shore hardness A 24/48/336 h | Viability, min. |
|---|---|---|---|---|---|---|
| 1 | U-30 M | 2.87 | 275 | - | 48/51/56 | 10 |
| 2 | U-30 M | 2.51 | 275 | - | 31/38/54 | 420 |
| 3 | AM-05 | 0.82 | 460 | 1.88 | -/-/- | 12 |
| 4 | AM-05 | 0.81 | 510 | 2.50 | -/-/- | 510 |

Data in Table 3.3 provide with a conclusion that manganese dioxide activity is supposed to mainly influence on the curing rate (viability), the Shore hardness A in 48 h and exerts almost no influence on deformation and strength properties and the final hardness value. Defective strength reducing structures do not form even at high curing rates (the viability is 10–12 min).

One must note some reduction of adhesion to duralumin for AM-05 sealant with high viability. It can be explained by improved conditions for contact on the sealant-duralumin boundary.

The curing rate is considerably influenced by the quantity of added manganese dioxide [23]. The analysis of $T_2$ variation during vulcanization (Fig. 3.10) shows, that the increase of curing agent concentration leads to a faster variation of the transverse relaxation time because of optimal concentrations of vulcanizing agents (the excess factor n is 2 ,2.5 times more than stoichiometric value). Their increase promotes vulcanization and network defectiveness [11, 24]. The kinetic curves of $T_2$ dependence on time have two vulcanization periods when vulcanizing agent concentrations are below optimal (n<2.5): the first period is the fast decline of $T_2$ at the start of reaction and its straightening to plateau, the second period is further change of the $T_2$ time (the plateau ends), until reaction of polysulfide oligomer vulcanization finishes and $T_2$ stays constant. Two vulcanization periods, observed on Fig. 3.10 at vulcanizing agent concentrations below optimal (n<2.5), can be caused by the presence of $Mn^{4+}$ ions with

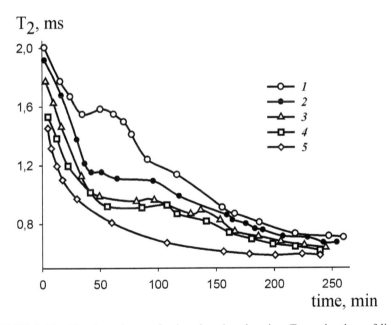

**FIGURE 3.10** The dependence of spin-spin relaxation time $T_2$ on the time of liquid thiokol vulcanization by manganese dioxide. The coefficient of vulcanizing agent excess above stoichiometry is: 1 – 1.5; 2 – 2.0; 3 – 2.5; 4 – 3.0; 5 – 4.0. Various mobility as it is shown above (Tables 3.1 and 3.2; Fig. 3.1). Variations in mobility can be explained by symmetry distortion in a local internal crystalline field in initial $MnO_2$.

The first vulcanization stage is faster and involves more mobile $Mn^{4+}$ ions of a vulcanizing agent (the narrow EPR spectral component), less stronger bound by a lig and crystalline field. Such $Mn^{4+}$ ions are considerably declined from octahedral structure and are less covalently bound to six neighboring oxygen atoms. Therefore, reactive oligomer groups are rapidly oxidized at the first stage. Then, as mobile $Mn^{4+}$ is consumed, other $Mn^{4+}$ ions start to participate in vulcanization process. These ions are stronger bound with neighboring oxygen atoms and are components of larger structures, so PSO vulcanization slows down.

As the concentration of a vulcanizing agent increases, the content of more mobile $Mn^{4+}$ ions grows. The second area of curves degenerates at the optimal $MnO_2$ concentration, corresponding to a double excess of stoichiometric value. Vulcanization process is in this case described by one smooth kinetic curve. It is significant, that the optimal $MnO_2$ concentration results in the most effective oligomer structuring, when the network chain density is maximal [1, 3].

Some commercial $MnO_2$ batches contain $Mn^{4+}$ ions with distinct differences in mobility and localization, but sometimes their structure proves to be more uniform (Table 3.2). In the last case, there are no two periods in polysulfide oligomer vulcanization (Fig. 3.11), and vulcanization reaction is described by one smooth kinetic curve for all curing agent concentrations [25].

It seems, that if localization of $Mn^{4+}$ ions in a vulcanizing agent is uniform enough, then PSO is oxidized in a mechanism proposed in [1, 18]. According to concepts in these publications, adsorption-desorption processes are crucial for PSO oxidation and mercaptide bonds are almost absent in vulcanizates.

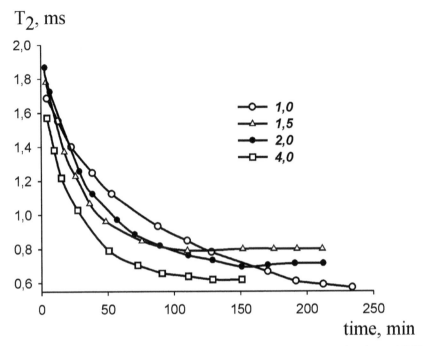

**FIGURE 3.11** The dependence of spin-spin relaxation time $T_2$ on the time of PSO vulcanization by homogeneous manganese dioxide.

When a vulcanizing agent has areas, varied in type of $Mn^{4+}$ localization and sizes and therefore, varied in mobility and structure, than PSO vulcanization mechanism can be described by following equations:

$$2\text{~R-SH} + MeO_2 \longrightarrow \text{~R-S-S-R~} + MeO + H_2O$$

$$2\text{~R-SH} + MeO \longrightarrow \text{~R-S-Me-S-R~} + H_2O$$

At first, mercaptide bonds form, then they are either oxidized by the excess of $MnO_2$ resulting in the occurrence of two vulcanization periods, especially at the curing agent concentration below the optimal value.

$$\text{~R-S-Me-S-R~} + MeO_2 \longrightarrow \text{~R-S-S-R~} + 2MeO$$

or eliminated at thermal vulcanization or by sulfur:

$$\sim\text{R-S-Me-S-R}\sim \xrightarrow{\;t_0\;} \sim\text{R-S-R}\sim + \text{MeS}$$

$$\sim\text{R-S-Me-S-R}\sim \xrightarrow{\;S\;} \sim\text{R-S-S-R}\sim + \text{MeS}$$

Both mechanisms seem to take place in reality, and their contribution is determined by the structure of initial vulcanizing agent.

## 3.2 THE MECHANISM OF POLYSULFIDE OLIGOMER VULCANIZATION BY SODIUM BICHROMATE

Sodium bichromate is one of the most active oxidative vulcanizing agents for PSO. Its application provides high curing rate and good strength properties. Sodium dichromate can also be easily converted to homogeneous state by dissolution in water or various solvents. These facts stipulate its application on industrial scale in real commercial compositions such as VITEF sealant. Therefore, it is interesting to study kinetic behavior of PSO vulcanization and to determine its mechanism.

Kinetic law patterns of vulcanization have been studied on the change of NMR ($^1$H) absorption line width, which grows along the process, reaching a distinct constant value.

Figure 3.12 demonstrates dH dependence on the quantity of sodium bichromate (SB) in a composition. PSO vulcanization is usually carried out at the excess of a curing agent to provide a full conversion of thiol groups. It has been determined earlier [1, 19, 26], that the optimal excess of SB is 1.5–2.0 plus equimolar, thus, compositions with the excess factor of 1–2.5 have been selected for research.

It can be derived from the Fig. 3.13, that vulcanization rate logically grows with the increase of SB excess. This effect can be explained by easier access to oligomeric thiol groups with the increase of a curing agent concentration and more frequent interaction acts. However, other interactions seem to occur in a system in addition to oxidation reaction, or SB is ineffectively used for vulcanization. Table 3.4 represents data on the influence of SB content on the density of chemical chains and the efficiency of chemical structuring of PSO vulcanizate. Provided data show extremism of the density of transversal chemical bonds and the effective chain density in vulcanizate, prepared with various excesses of SB.

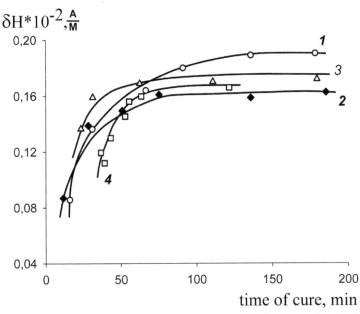

**FIGURE 3.12** The dependence of NMR absorption line width on vulcanization time for compositions with varied content of SB and the excess factor of: 1–1.0; 2–1.5; 3–2.0; 4–2.5.

Comparison of theoretical and experimental densities of chemical structuring shows, that the efficiency of chemical structuring does not exceed 50% of theoretical value even at the curing agent excess factor n=3.0. If SB quantity is close to equimolar, the efficiency of chemical structuring is 10–20% of theoretically possible.

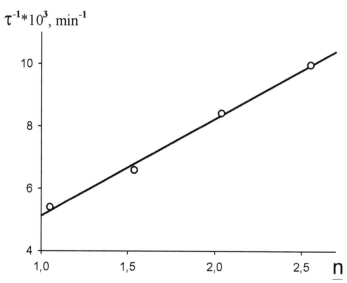

**FIGURE 3.13**   The dependence of vulcanization reaction rate on the quantity of SB (n is the excess factor).

**TABLE 3.4**   The Influence of Sodium Bichromate Content on the Efficiency of Chemical Structuring of Liquid Thiokol Vulcanizate

| Excess factor | Effective density of transversal bonds, $10^4$ mol/sm$^3$ | The density of chemical bonds, $10^4$ mol/sm$^3$ | | Chemical structuring efficiency, % |
|---|---|---|---|---|
| | | Theory | Experiment | |
| 1.0 | 2.34 | 1.15 | 0.13 | 11.3 |
| 1.5 | 3.26 | 1.73 | 0.43 | 24.8 |
| 2.0 | 3.30 | 2.30 | 1.38 | 60.0 |
| 2.5 | 3.0 | 2.88 | 1.40 | 48.6 |
| 3.0 | 2.80 | 3.46 | 1.60 | 46.2 |
| 4.0 | 2.60 | 4.61 | 0.90 | 19.5 |

When the excess factor of a curing agent increases, two factors start to exert influence on structuring: the increase of structural efficiency due to the growth of vulcanizing agent concentration and the enhancement of structural defects due to the increase of vulcanization rate. When the content of a curing agent is small and below the optimal value, the increase of SB concentration is the prevailing factor. When the content of a vulcanizing agent is above optimal, a substantial increase of vulcanization rate prevails, resulting in formation of a defective network.

Some aspects of PSO vulcanization by sodium bichromate have been studied for interpretation of extreme behavior of the effective transversal chain density and structuring efficiency parameters.

There were suggestions earlier [27, 28], concerning formation of coordinative bonds of $Cr^{3+}$ or $Cr^{5+}$ ions with oxygen or sulfur electron donors on PSO polymer chain [27, 28]. PSO vulcanization process is supposed to occur in following chemical reactions:

$$Na_2Cr_2O_7 \xrightarrow{HOH} 2Na^+ + Cr_2O_7^{2-}$$
$$Cr^{6+} + 2e \longrightarrow Cr^{4+}$$
$$Cr^{4+} + Cr^{6+} \longrightarrow 2Cr^{5+}$$
$$Cr^{5+} + 2e \longrightarrow Cr^{3+}$$

$Cr^{3+}$ ion or chromium with other oxidation number forming after the reduction of thiol PSO groups by sodium bichromate can enter into in complexation reactions.

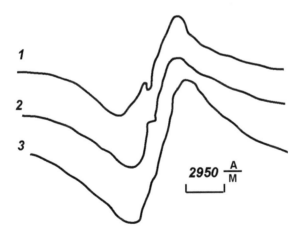

**FIGURE 3.14** EPR spectra, taken during liquid thiokol vulcanization by SB (time from the start of vulcanization: 1 – 15 min.; 2 – 30 min.; 3 – 40 min.).

Figure 3.14 shows EPR spectra taken in radio spectrometer resonator during PSO vulcanization by SB. They are two-component lines indicating the presence of chromium atoms with various oxidation numbers.

A narrow spectral component extracted to a separate figure (Fig. 3.15), indicates Cr$^{5+}$complexes by its g-factor value, adsorption line shape and HFS constant value. Although odd isotope Cr$^{53}$ has a magnetic moment of 0,47354, its spin is relatively high (3/2). Thus, there are at least two the most remote lines are allowed [7] (g-factor value for the narrow component is 1.978 + 0.002, A$_{11}$ 10$^4$ ~30 sm$^{-1}$).According to [29, 30], the wide spectral line can indicate the presence of Cr$^{3+}$ ions in a system and the width of this line (~480e) is stipulated by g-factor anisotropy.

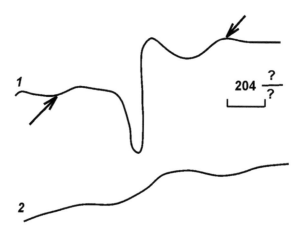

**FIGURE 3.15**  The narrow EPR spectral component shape (hyperfine structure is marked by arrows). Reaction duration: 1 – 7 min.; 2 – 22 min.

The narrow spectral component disappears in 15–20 min after vulcanization start. Meanwhile, the broad one remains in entire vulcanization process and its integral intensiveness grows along this process. This effect can be explained by either transition of Cr$^{5+}$ complexes to Cr$^{3+}$ complexes or formation of Cr-O-Cr bridges, where bound chromium is diamagnetic [31]. According to literary data, [32], Cr$^{5+}$ complexes are mainly tetrahedral and their ligands are usually oxygen atoms. It can be one of the factors, determining high effective structuring density of vulcanizate.

At the start of vulcanization, when a system is highly mobile, Cr$^{5+}$ ions can enter into complexation with oxygen atoms, forming in a reaction. It should exert influence on the completeness of oligomeric HS-group conversion, when the PSO:SB ratio is equimolar. Indeed, the efficiency of chemical structuring of vulcanizate is only 11, 3% in this case (Table 3.4). Therefore, it is necessary to take the SB in excess at PSO.

Coordination of forming $Cr^{3+}$ complexes is rather difficult to determine, as the broad EPR spectral component (dH = 480e, g =1.976 + 0.03) gives little information on ligands of these complexes. However, according to Refs. [33–35], EPR line width increases substantially for $Cr^{3+}$ complexes in sulfur neighborhood or in a matrix. It cannot be observed in the presence of other ligands.

PSO chain in the studied system contains many disulfide bridges, being able to enter into a complexation reaction. These sulfur atoms can coordinate in first or second coordination sphere.

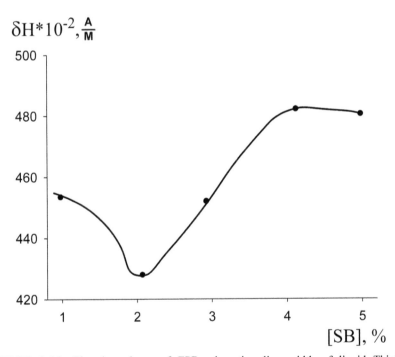

**FIGURE 3.16** The dependence of EPR adsorption line width of liquid Thiokol vulcanizate on the quantity of SB.

Figure 3.16 represents the dependence of EPR absorption line width on vulcanizing agent excess factor. This dependence confirms the fact, that minimal quantity of coordination bonds is observed at n=2. The value of n=2 seem to favor the most effective mercaptan-to-disulfide group oxidation.

Narrowing of EPR adsorption line width from 456 Gs to 18, 5 Gs at the transition from usual sealants to vulcanizate and 170-hour water extraction also confirms the presence of additional coordinative bonds in vulcanizate. These bonds can break, when sealant is exposed to a polar medium.

There were some suggestions on PSO vulcanization mechanism, based on the study of NMR line width temperature dependencies, taken for liquid Thiokol vulcanizate, cured by various quantities of SB. Figure 3.17 demonstrates NMR line width temperature dependences, taken for liquid Thiokol vulcanizate, cured by SB. All the samples have rapid narrowing of the line width dH in a temperature range of −60: −0°C, while the variation of vulcanizing agent concentration (1.0<n < 3.0) exerts little influence on the temperature dependence of a line width. Activation energy values of the analyzed motion types in vulcanizate, calculated of NMR spectra temperature dependences, is 27 kJ/mol for n=1.0 and 32.8 kJ/mol for n = 3.0.

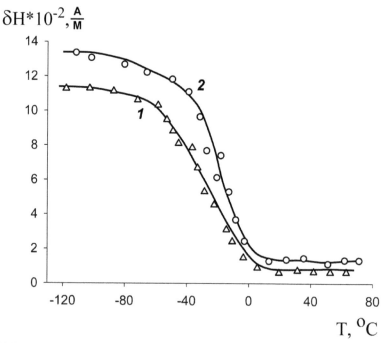

**FIGURE 3.17**   The temperature dependence of NMR line width on SB excess factor for liquid thiokol vulcanizates: 1 – n=1.0; 2 – n=3.0.

NMR spectra of PSO vulcanizate, cured by SB, are two-component curves in a certain temperature range (Fig. 3.18). These vulcanizate have spectral line with a complex shape in a more narrow temperature range (from −15 to −50°C) and demonstrate clearly a concentration dependence on the quantity of added SB. As it's shown above, the most intensive process at n=2.0 is oligomeric thiol group oxidation, while additional system structuring by complexation is less visible (Fig. 3.16). Therefore, a complex NMR line shape for such a quantity of a vulcanizing agent shows up unclearly, while the two-component NMR line is well visible when the excess of a vulcanizing agent is increased up to n=4,0. We can suppose, that appearance of narrow

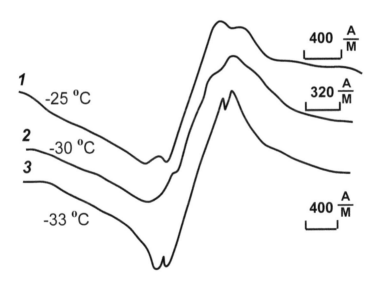

**FIGURE 3.18**   NMR spectra, taken for SB cured liquid thiokol vulcanizates with varied excess factor: 1 – 1.0; 2 – 2.0; 3 – 4.0.

NMR spectral component for SB cured PSO vulcanizates is the effect of additional complexation in a system, which contributes to a polymer network density.

This component seems to disappear, when temperature is further decreased, because more mobile chain components, created by these bonds, become identical with polysulfide oligomer vulcanizate's main chains in their molecular motion. Therefore, PSO oxidation results in formation of PSO complexes with $Cr^{5+}$ and $Cr^{3+}$ atoms of a vulcanizing agent. These atoms appear as the result of consecutive $Cr^{6+}$ oxidation in a system in the following reactions:

$$Cr^{6+} + 2e \longrightarrow Cr^{4+}$$
$$Cr^{4+} + Cr^{6+} \longrightarrow 2Cr^{5+}$$
$$Cr^{5+} + 2e \longrightarrow Cr^{3+}$$

A summarized equation for the process looks like this:

$$2\,Cr^{6+} + 2\,R\text{-}S \longrightarrow 2\,Cr^{5+} + R\text{-}S\text{-}S\text{-}SR$$
$$Cr^{4+} + 2\,R\text{-}S^{-} \longrightarrow Cr^{3+} + R\text{-}S\text{-}S\text{-}R$$

Obtained data confirm the mechanism of adsorption-desorption PSO oxidation by metal dioxides and explain high chemical structuring density of PSO vulcanizates by additional complexation in a system in the case of SB used as a vulcanizing agent. Complexes make a substantial contribution to the effective density of polymer chains. Consideration of this contribution provides a more precise estimation of SB content required for production of thiokol sealants with tailored properties.

## 3.3   LIQUID THIOKOL VULCANIZATION BY ZINC DIOXIDE

There is a firm demand for white (light) sealants in construction in recent years, in particular, for sealing of interpanel seams in house building. There are several production technologies for such sealants, which take requirements of "GOST 25621–83" (Polymer materials and products for sealing and packing used in construction. Classification and general technical specification) standard into consideration [36, 37].

The first method is the simplest: addition of titanium dioxide to seam sealants, as it is highly effective pigment. Existing sealants, such as AM-05, LT-1, SG-1, etc.,contain 3-4% of manganese dioxide as a vulcanizing agent and chalk as a filler and Therefore, are of dark-gray color. It is required to add at least 10–15 mass parts of titanium dioxide per 100 mass parts of PSO to make light-gray sealants and at least 30 weight parts of titanium dioxide to make white sealants. Such sealants have to be sold at much higher prices and their sales are Therefore, considerably limited.

The second way is PSO curing is copolymerization in the presence of reactive compounds with functional end groups, which are, in turn, reactive to SH groups: NCO-groups, epoxy groups, double bonds, etc. [3, 38–40]. However, when equimolar to PSO quantities of rigid chain epoxy resins (20–40%) are added, elastic and low-temperature properties may decrease. If oligomers with other structure are used, such as isocyanate compounds, one may expect narrowing of the processing temperature range, shortening of pre-use storage time etc.

It is possible to make white sealants, using organic peroxides for PSO curing. However, thermal resistance of such compositions does not exceed +75 °C, that can considerably reduce their durability, because sealant's surface can be heated by solar radiation up to the of +80–90°C in summer even in Russian moderate climate zone. Therefore, there are few vulcanizing agents for production of durable seam sealants, which are able to keep their sealing qualities in conditions of alternating deformations, UV, ozone and water effects in the temperature range of –40°C to +100 °C.

Zinc oxide and peroxide are known to be used for curing liquid Thiokol sealants [3, 40]. Application of zinc oxide can help in making white sealants with good deformation and strength properties various sulfur-containing compounds are used to promote curing in this case: mercapto benzothiazole, tetra alkyl thiuramdisulfides [40–44], thiuram-sulfur combinations [45]. Addition of light color fillers always leads to making of white sealants. Reactive sealants, such as glycidylpropyltriethoxysilane are used to increase adhesion [40, 42].

There was research on the synthesis of ZnO-cured Thiokol sealants with a prominent range of properties and no above disadvantages. Research has resulted in preparation of liquid Thiokol-based sealants with good strength and adhesive properties. They were cured by ZnO in the presence of E-40 Diane epoxy resin and activating composition, which combines thiuram and diphenylguanidine. Variation of sealant properties with the content of vulcanizing composition and E-40 resin has been determined to have same behavior, independently on strengthening filler type (titanium dioxide or carbon black) (Fig. 3.20) [6, 46, 47].

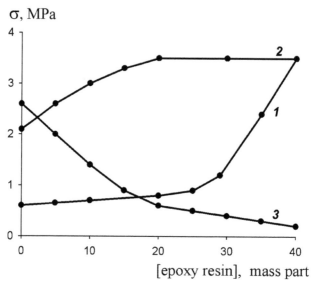

**FIGURE 3.19**   The dependence of conventional strength on epoxy resin content: 1 – ZnO = 5 mass parts; 2 – ZnO = 25 mass parts; 3 – MnO$_2$.

As Fig. 3.19 shows, the increase of epoxy resin content leads to different behavior of strength of thiokol sealants, prepared with various vulcanizing agents. If MnO$_2$ is used, strength decreases due to poor curing processes, formation of defective structures and increase of uncured resin and Thiokol in a sealant (according to sol-gel analysis data) [48]. If ZnO curing is carried out, the opposite effect is observed: the increase of epoxy resin content makes a sealant considerably more durable with maximum performance at Thiokol: resin ratio of 100: 30–40 mass parts, that approximately corresponds their equimolar ratio. Provided data (Fig. 3.19) also indicate that the increase of resin content effects drastically on a strength at reduced ZnO content ZnO (5 mass parts).

A composition of sealant has been optimized. A central flexible composition plan for four factors has been worked out for this purpose. A composition was optimized by factors, which exert the strongest influence on sealant's properties according to obtained data.

Adequate regression equations regarding estimated parameters have been obtained after implementation of a matrix, calculation of regression coefficients [49] and screening of insignificant coefficients:

• regarding conventional strength:

$$f_P(MPa) = 2,31 - 0,28x_1 + 0,72x_3 + 0,57x_4 + 0,075x_2^2 - 0,076x_3^2 +$$
$$+ 0,1x_1x_3 + 0,128x_2x_3 - 0,283x_3x_4$$

- regarding relative elongation:

$$\varepsilon_{RELATIVE}(\%) = 312 - 54x_3 - 73x_4 + 15x_3 + 46x_3x_4$$

- regarding adhesion to duralumin at delamination:

$$A(kN/m) = 2,50 - 0,24x_3 - 0,27x_3 - 0,32x_3x_4$$

As strength-regarding regression equation shows, ZnO and E-40 resin contents exert the deepest influence on this parameter (Fig. 3.20).

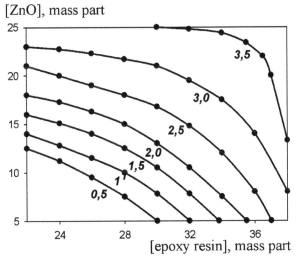

**FIGURE 3.20**   The dependence of strength (MPa) on ZnO and E-40 content.

Indeed, the influence of these factors on a strength level is approximately the same. High strength level can be achieved at a minimal content of ZnO and maximal content of resin and vice versa. This effects are also confirmed by Fig. 3.21(a–c).

Thiuram exerts a distinct effect on strength only at ZnO content of 25 mass parts (Fig. 3.21c). More resin added to a sealant does not increase strength at the maximum ZnO content. This effect can be explained by stronger influence of increased ZnO additives on a curing process.

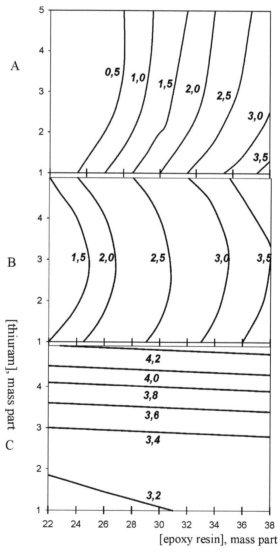

**FIGURE 3.21**    The dependence of strength (MPa) on thiuram and E-40 content of: (a) 5 mass parts; (b) 15 mass parts; (c) 25 mass parts of ZnO.

When more E-40 and ZnO added to compositions (Fig. 3.22), relative elongation value reduces considerably. ZnO and E-40 effect is approximately the same in studied boundaries that once again confirms our proposal of E-40 being a vulcanizing agent, as ZnO is.

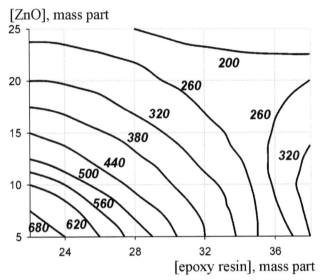

**FIGURE 3.22**    The dependence of relative elongation on ZnO and E-40 content.

Extreme curing agent dosage (E-40 content > 35 mass parts) leads to further decrease of relative elongation. It was not observed for ZnO. The analysis of obtained experimental data (Fig. 3.23) confirms active participation of E-40 in PSO curing, while its studied dosages do not plasticize a composition. In addition, E-40 activity is similar to that of zinc dioxide.

The study of ZnO and E-40 influence on adhesion strength (Fig. 3.23) reveals complex effects. Adhesion properties increase with E-40 content of up to 15 mass parts in ZnO-containing compositions. Excess of ZnO reduces adhesion with the increase of E-40 content. The prevailing process here is the adsorption of epoxy resin hydroxyl groups on ZnO surface and their corresponding screening [50].

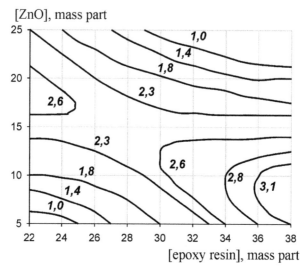

**FIGURE 3.23** The dependence of adhesion to duralumin to zinc oxide on E-40 content.

Therefore, obtained results advance a proposal of E-40 resin–Thiokol interaction involving thiokol SH-groups, when this resin is added to compositions cured by ZnO. The degree of participation of epoxy resin in curing is determined by the content of ZnO.

ZnO and E-40 participation in curing of liquid thiokol can be described by the following reactions:

$$1 \quad \sim R\text{-}SH + CH_2\text{-}CH\text{-}R'\sim \longrightarrow \sim R\text{-}S\text{-}CH_2\text{-}CH\text{-}R'\sim$$
$$\underset{O}{\diagdown\diagup} \qquad\qquad\qquad \underset{OH}{|}$$

$$2 \quad 2\sim R\text{-}SH + ZnO \longrightarrow \sim R\text{-}S\text{-}Zn\text{-}S\text{-}R\sim + H_2O$$

Curing process is activated by aminous promoters, such as thiuram and diphenylguanidine. Both reactions seem to take place in studied systems. However, the first reaction prevails at minimal ZnO content and vice versa.

Kinetic regularities of curing of thiokol sealants with various composition have been studied to obtain a clearer picture of a curing mechanism for thiokol sealants with ZnO and E-40 curing agents, as well as to estimate their performance (Fig. 3.24a, 3.24b) and the influence of a formulation on sealant's properties (Table 3.5). Properties were estimated for sealants, cured for 14 days at 70°C (samples 1–4) and for one day at 70°C (samples 5–6).

As Fig. 3.24a and 3.24b show, compositions, which contain, in addition to ZnO, epoxy resin, demonstrate high vulcanization rate and acquire good strength properties in 24 h already. Sealants with only ZnO additions demonstrate drop of relative elongation after storage at 70°C for 5 days. Sealants were aged for 14 days at 70°C and then exposed to extraction by hot toluene with further determination of sol-fraction composition and sulfur content. Sol-gel analysis data are given in Table 3.5.

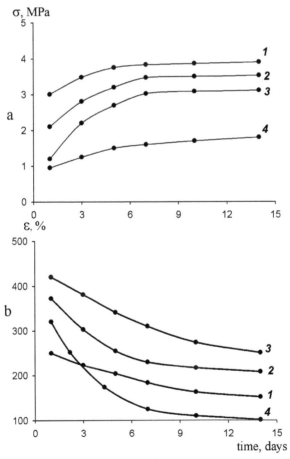

**FIGURE 3.24** The dependence of strength (a) and relative elongation (b) on sealant's curing time (T = 70°C): 1–sample 1; 2–sample 2; 3–sample 3; 4–sample 4 (Table 3.5).

Sol fraction composition has been determined for each formulation, considering the content of sulfur.

Obtained data prove, that sol-fraction compositions of each studied formulation are similar, as they mainly include liquid thiokol. Epoxy resin is bound (more than 80%) in a gel-fraction, probably forming a copolymer. The results of sol-gel analysis made for U-30 MES-10 sealant, do not contradict to earlier data [48]. They just confirm the necessity of increasing the curing time for epoxy resin-containing thiokol sealants, cured by manganese dioxide. In addition, rapidly cured composition 5 with no epoxy resin (Table 3.5) fully dissolves in toluene, while a composition with epoxy resin does not.

Considering above facts, we can conclude, that the reaction 1 is more rapid and preferable for compositions with both ZnO and E-40 resin. Considering, that liquid thiokol is present in sol-fraction due to breaking of sealant mercaptide bonds, ZnO is a vulcanizing agent too.

Curing kinetics of unfilled vulcanizates has been studied using infrared spectroscopy (IRS) method at 70°C. As a result, zinc dimethyldithiocarbamate was determined to appear at the initial moment. Zinc dimethyldithiocarbamate formation of thiuram is a known fact. It occurs in the following reaction at the presence of thiokol [51]:

$$3 \quad (CH_3)_2N-\underset{\underset{S}{\|}}{C}-S-S-\underset{\underset{S}{\|}}{C}-N(CH_3)_2 + ZnO + 2\,R\text{-}SH \longrightarrow$$

$$\longrightarrow [(CH_3)_2N-\underset{\underset{S}{\|}}{C}-S]_2Zn + R\text{-}S\text{-}S\text{-}R + H_2O$$

Such a mechanism is possible. However, it takes a small part of the process, as it would lead to formation of a sealant, being fully insoluble in toluene.

In addition, IRS data report formation of carbonyl groups during curing. A carbonyl group can form at increased curing temperature, while the presence of ZnO favors free-radical opening of both thiuram's and liquid thiokol's disulfide bonds forming zinc dimethyldithiocarbamate in the first case. Epoxy resin can interact with thiokol in a following reaction:

$$4 \quad \sim R\text{-}S\text{-}S\text{-}R\sim + ZnO \longrightarrow \sim 2R\text{-}S^{\bullet} + \underset{O}{\overset{\displaystyle CH_2-CH\sim R'}{\diagdown\!\diagup}} \longrightarrow \sim R\text{-}S\text{-}CH_2\text{-}\underset{\underset{O}{\|}}{C}H\text{-}R'\sim$$

Mercaptide bond seems to transform into monosulfide bond after cutting of sealants by ZnO at prolonged heating (7–14 days, 70°C) in the following reaction:

**TABLE 3.5** The Composition and Properties of Thiokol Sealants

| Composition | 1 | 2 | 3 | 4 | 5 | 6 |
|---|---|---|---|---|---|---|
| NVB-2 thiokol | 100 | 100 | 100 | 100 | 100 | 100 |
| P-803 carbon black | 30 | 30 | 30 | 30 | ¾ | ¾ |
| E-40 Epoxy resin | 35 | 35 | 13 | ¾ | ¾ | 35 |
| ZnO | 30 | 5 | ¾ | 30 | 30 | 30 |
| Paste number 9 | ¾ | ¾ | 20 | ¾ | ¾ | ¾ |
| Thiuram «D» | 2 | 2 | ¾ | 2 | 2 | 2 |
| Diphenylguanidine | 2 | 2 | 2 | ¾ | 2 | 2 |
| Properties: | | | | | | |
| Conventional strength at rupture, MPa | 4.09 | 3.58 | 1.62 | 3.11 | ¾ | ¾ |
| Relative elongation at rupture, % | 220 | 220 | 338 | 155 | ¾ | ¾ |
| Residual elongation at rupture, % | 2 | 2 | 2.5 | 1.5 | ¾ | ¾ |
| Adhesion to D-16 duralumin, kN/m | 3.45 | 4.60 | 1.64 | 1.30 | ¾ | ¾ |
| Sol fraction content per sealant, % | 22.5 | 23.8 | 21.9 | 23.5 | ¾ | ¾ |
| Sol fraction content per organic phase, % | 32.2 | 29.8 | 29.2 | 37.3 | 100 | 45.7 |
| Sulfur content in a sol fraction, % | 32.4 | 31.7 | 26.3 | 40 | ¾ | ¾ |
| Thiokol content in a sol fraction,%of a total quantity | 34.3 | 31.2 | 24.7 | 36.1 | ¾ | ¾ |
| Epoxy resin content in a sol fraction,%of a total quantity | 17.1 | 17.1 | 7.0 | ¾ | ¾ | ¾ |

$$5 \quad \sim\!R\text{-}S\text{-}Zn\text{-}S\text{-}R\!\sim \xrightarrow{t,\,°C} \sim\!R\text{-}S\text{-}R\!\sim \; + \; ZnS$$

$$6 \qquad or \quad \xrightarrow{S} \sim\!R\text{-}S\text{-}S\text{-}R\!\sim \; + \; ZnS$$

Both these variants are possible, as free sulfur is always present in thiokols. However, sulfur content is insufficient and the reaction 5 is the main way to get ZnO-cured sealants, insoluble in toluene. Zn–S bond is much stronger, than Pb–S bond, so mercaptide-to-monosulfide bond transformation can be carried out only by means of heating and additional sulfur.

We can conclude that several concurring reactions can occur in ZnO-cured sealants in the presence of epoxy resin. First of all, liquid thiokol interacts with epoxy resin. This reaction is catalyzed by ZnO-thiuram interaction products, forming polymer structures. Maximum strength is achieved at equimolar ratio of epoxy and sulfhy dryl groups. Produced sealants have a range of properties overcoming that of commercial formulations with epoxy resin (U-30 MES-10). They possess the property of strong, stable and heating-free adhesion and can be used for the same purposes.

It is easy in this case to get both white and colored sealants; Therefore, these products can be successfully used in construction.

## 3.4   THE INFLUENCE OF VULCANIZING AGENT NATURE ON THE PROPERTIES OF TPM-2 POLYMER SEALANT

End properties of sealants and their operating performance are mainly determined by the nature of a vulcanizing agent, its activity, possibility of its participation in complexation, the character of forming transversal bonds, the degree of influence of its reduction products on properties. The influence of vulcanizing agent's nature on properties of Thiokol sealants is described in Chapter 2. A comparative analysis of TPM-2-based sealants, cured by manganese or zinc dioxides, has shown that they are similar in their strength and adhesive properties. Application of manganese dioxide provides sealants with higher deformability [47, 52]. Thermo mechanical and relaxation properties of vulcanizate networks made using manganese or zinc dioxide have been studied to establish their contribution to the properties of TPM-2-based sealants [52, 53].

Thermomechanical analysis (TMA) data prove that formulations with zinc dioxide have less wide rubber-like-elasticity interval, and their viscous-flow transition temperature ($T_{flow}$) 20–25°C less, than that of sealants, cured by manganese dioxide (Fig. 3.25).

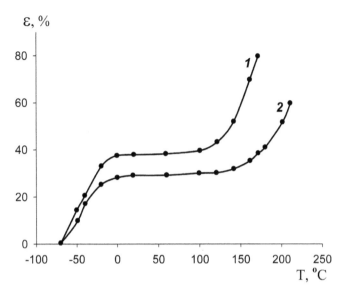

**FIGURE 3.25**   TMA curves of ZnO (1) and MnO (2) cured sealants.

A glass-transition temperature ($T_{glass}$) is the same in both cases and stays in a temperature range of −60 +2°C. Differential-thermal analysis (DTA) has determined the same $T_{glass}$ value for these compositions.

Mercaptide bonds, forming in thiokol sealants as a result of SH-group oxidation by metal oxides and dioxides during vulcanization, are known to be insufficiently stable, as they can transform into mono- and–disulfide bonds during thermal processing [54]. To study these processes, TMA has been carried out for sealants with ZnO or $MnO_2$. These sealants have been thermally processed before curing at 150°C for 1–2 h. It has been determined, that a formulation with ZnO loses almost fully its elastic properties ($T_{flow}$ = 25–40°C), whereas formulations with $MnO_2$ have a broader rubber-like elasticity "plateau"as well as the increase of $T_{flow}$. Therefore, processing temperature of compositions with ZnO has been reduced down to 140°C.

TMA of sealants, exposed to thermal treatment of various duration have discovered that the dependence of $T_{flow}$ on time has the shape of sloping extremum for both systems, while their deformability is similar at a room temperature and inclined to a steady growth (Fig. 3.26).

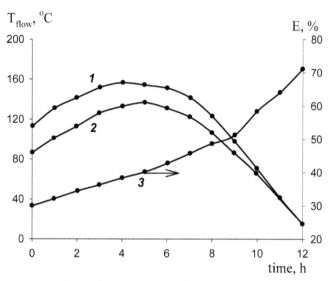

**FIGURE 3.26**   The effect of thermal treatment duration on the viscous flow temperature $(T_{flow})$–1, 2 and deformability –3 of sealants.

Such a behavior of $T_{flow}$ dependences seem to be the result of intensi-fication of condensation, cross-linking and transversal chain destruction concurrent processes at 140–150°C, while destruction processes prevail after 5 h of heating and lead to a definite loss of elastic properties. Air oxy-gen oxidation seems to become more important at the stage of prevailing destructive processes. These processes become visible on DTA curves at temperatures above 200°C (Fig. 3.27).

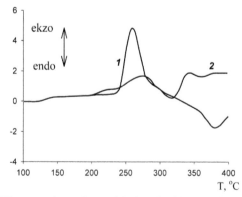

**FIGURE 3.27**   DTA curves for sealants with zinc dioxide (1) and manganese dioxide (2).

A various style of heat generation attracts attention here. It indicates differences in the chemistry of occurring processes. The composition with zinc dioxide shows a sharp exothermic peak, while the composition with manganese dioxide has considerably reduced intensity of heat generation. As the composition of sealants is identical, DTA differences indicate different curing agents. Exo-and-endothermic effects can be first of all explained by main chain destruction, involving mercaptide and disulfide bonds as well as by destruction of "salt" bonds, which can forms in sealants with zinc dioxide [52].

The influence of oxidant nature on the structure of forming network can be seen after comparison of relaxation characteristics of systems. For example, compositions with manganese dioxide relax faster and more completely, than compositions with zinc dioxide after compressive deformation (= 50%, holding period = 1 h) (Fig. 3.28).

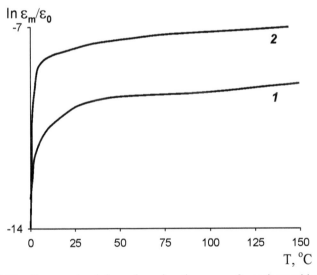

**FIGURE 3.28**  Compressive deformation relaxation curves for sealants with zinc (1) and manganese (2) dioxides.

Differences in the structure of sealants also influence their frost resistance. Formulations with zinc dioxide have lower elastic recovery factor. Its value at −60°C is 0.13, while $K_r = 0.38$ for formulations with manganese dioxide.

Therefore, research has been carried out and sealants cured by zinc dioxide have been determined to have lower thermal and relaxation properties, than sealants cured by manganese dioxide because of various structure of forming vulcanization network.

## KEYWORDS

- mechanisms
- oligomers
- polysulfide
- properties of sealant
- structure
- vulcanization

## REFERENCES

1. Averko-Antonovich, L. A., Kirpichnikov, P. A. (1978). Materials of International Conference on Caoutchoucand Rubber (in Russian). Kiev. 44–51.
2. Pratt, E. F., McCowera, T. P. (1964). J. Org. Chem.V.29. 1540–1543.
3. Averko-Antonovich, L. A., Kirpichnikov, P. A., & Smyslova, R. A. (1983). Polysulide oligomers & Related Sealants (in Russian). Leningrad: Himija −128 p.
4. Minkin, V. S. (1997) NMR in Commercial Polysulfide Oligomers (in Russian). Kazan. "ABAK" 222 p.
5. Tavrin, A. E., Tejtel'baum, B. Ya. (1969). Dokl. AN SSSR (in Russian) 186(5), 1065–1068.
6. Al'tshuler, S. A., Kozyrev, B. M. (1972). EPR of Compounds, Formed by Elements from Interjacent Groups (in Russian). Moscow: Nauka 262 p.
7. Kuska, H., Rodzhers, M. (1970). EPR of Transition Metal Complexes (in Russian). Moscow. Nauka, 82 p.
8. Minkin, V. S., Liakumovich, A. G., & Khakimullin, Yu. N. (1997) et al. Abstracts of 6th International Conference on Chemistry & Physical Chemistry of oligomer (in Russian). Kazan. Vol.1 P. 297.
9. Minkin, V. S., Khakimullin, Yu. N., Minkina, Ju. V., & Chistjakov, V. A. (1999). Abstracts of 5th International Conference on Intensification of Petrochemical Processes "Neftehimija 99" (in Russian). Nizhnekamsk. Vol.2. P. 63–64.
10. Nikolaeva, I. V., & Tihomirova, N. N. (1966) Zhurnalstrukturnojhimii (in Russian). Vol. 7(3), 351–354.
11. Bogomolova, L. D., Lazurkin, V. N., Chepeleva, N. V., & Bel'skaja, A. A. (1965) DAN SSSR (in Russian), 165(6), 1336–1338.

12. Rogger, J. F., Rogger, M. T., & Amer.Chem, J. (1959) Soc., 81, 1849–1852.
13. Panfilov, V. V., Verewagin, A. F., & Doklady (1964) AN SSSR (in Russian). Vol. 154(4), 819–824.
14. Gastner, I., Newell, I. S., Holton, W. S., & Slichter, C. P. J. (1960), Chem. Phys., v.32, 3, 668–672.
15. Minkin, V. S. (1975) Chemical Sciences Candidate's Dissertation (in Russian), Kazan, KHTI.
16. Maklakov, A. I., & Pimenov, G. G. (1968). Vysokomolek. Soed. (in Russian). Series A. Vol.10(4), 662–667.
17. Kontos, E. G., & Slichter, C. P. (1962). J. Polym. Sci., v.61, 61–64
18. Nefed'ev, E. S. (1979). Chemical Sciences Candidate's Dissertation (in Russian). Kazan. 169 p.
19. Averko-Antonovich, L. A., & Muhutdinova, T. Z. (1975). The Study of Liquid Thiokol Vulcanization by Sodium Bichromate by Stress Relaxation Method-Kazan. 12 p. Deposited in ONIITJeHIM (Cherkassy) 06.13.1975, 1730/75.
20. Minkin, V. S., Averko-Antonovich, L. A., & Kirpichnikov, P. A. (1973).Vysokomolek. Soed. (in Russian). SeriesB. Vol.15(1),24–26.
21. Minkin, V. S., Khakimullin, Y. N., Idiatova, A. A., Minkina, Y. V., Deberdeev, R. Ya., & Zaikov, G. E. (1999). Russian Polymer News. Vol.4(3), 13–15
22. Khakimullin, Y. N., Minkin, V. S., Idiatova, A. A., Minkina, Y. V., Deberdeev, R. Ya., & Zaikov, G. E. (2000). Inter. I. Polym. Matter. Vol.47, 373–378.
23. Minkin, V. S., Khakimullin, Y. N., Minkina, Y. V., Chistyakov, V. A., Deberdeev, R. Ya., & Zaikov, G. E. (1999). Russian Polymer News.Vol.4(4), 1–3.
24. Minkin, V. S., Averko-Antonovich, L. A., & Kirpichnikov, P. A. (1983). Izv. VUZov. "Chemistry & Chemical Technology" series (in Russian). Vol 26(3), 348–351.
25. Khakimullin, Y. N., Minkin, V. S., Minkina, Y. V., Chistyakov, V. A., Deberdeev, R. Ya., & Zaikov, G. E. (2000). Inter. I. Polym. Matter Vol 47, 367–372.
26. Muhutdinova, T. Z., Averko-Antonovich, L. A., Naumova, G. V., & Kirpichnikov, P. A. (1971). KSTU Transactions (in Russian), Issue 46, 79–80.
27. Averko-Antonovich, L. A., Rubanov, V. I., & Klimova, L. I. (1977) Vysokomol. soed. (in Russian), Series A, Vol. 19(7), 1593–1598.
28. Averko-Antonovich, L. A., Muhutdinova, T. Z., & Minkin, V. S. (1974) Jastrebov. Vysokomol. Soed. (in Russian), Series A, Vol. 16(8), 1709–1713.
29. Garif'janov, N. S. (1955). Doklady. AN SSSR (in Russian), Vol. 103(1), 46–49.
30. Garif'janov, N. S. (1959). Zhournal experimental noiiteoreticheskoifiziki (in Russian) Vol.37(6), 12–15.
31. Remi, T. (1966). Inorganic Chemistry (in Russian), Moscow, Mir, 128.
32. Garif'janov, N. S. (1962). Solid Body Physics (in Russian), Vol. 4, 2450–2454.
33. Locker, D. R., Gale, K. A., Kulp, B. A., & Dorain, P. B. (1966). Bull. Amer. Phys. Soc., 11(6), 719–722.
34. Auzing, P., Orton, I. W., Wertz, I. E. (1962). Intern. Conf. Paramagnetic Resonance, Jerusalem, 90–92.
35. Rivkind, A. I. (1959). Abstracts of the National EPR Conference (in Russian), Kazan, 12–13.

36. Smyslova, R. A., Shvec, V. M., & Sarishvilli, I. G. (1991). Application of Curable Sealants in Construction Technics. A Review (in Russian), VNIINTIJe PSM, Series 6, 2, 30 p.
37. Hozin, V. G. (1997). Transactions of theoretical and practical conference. "Production & Consumption of Sealants and Other Construction Compositions: Present State & Prospectives." (In Russian) Kazan. 9–20.
38. Li, T., S. P. (1995). Kauchukirezina (in Russian) 2, 9–13.
39. Lucke, H. (1994). Aliphatic Polysulfides. Monograph of an elastomer. Publisher Huthig & Wepf Basel, Heidelberg, New York. 191 p.
40. Smyslova, R. A. (1984). Liquid Thiokol Sealants (in Russian).Moscow: CNIIT Jenefte him, 67 p.
41. Patent 3923754 USA, MKI C 08 G 75/00.
42. Patent USA4314920, MKI S08 L 93/00.
43. Gafurov, F. Sh., & Khakimullin, Yu. N. (1995). Abstracts of 2nd Russian Theoretical & Practical Conference on Rubber "Feedstock and resources for Rubber Industry" (in Russian). p.154.
44. Idijatova, A. A., Khakimullin, Yu. N., Gafurov, F. Sh., & Liakumovich, A.G. (2000). Abstracts of 7th Russian Theoretical & Practical Conference on Rubber "Feedstock and Resources for Rubber Industry" (in Russian). 289–290
45. Patent 2000752 Germany, MKI G 02 B 21/36.
46. Idijatova, A. A., Khakimullin, Yu. N., Gafurov, F. Sh., & Liakumovich, A. G., Kauchukirezina (2002) (in Russian).4 P. 25–29.
47. Idijatova, A. A. (1999).Chemical Sciences Candidate's Dissertation (in Russian), Kazan, KSTU.
48. Muhutdinova, T. Z., Shahmaeva, A. K., Gabdrahmanov, F. G., & Satarova, V. M. (1980). Kauchukirezina (in Russian), 1, 12–15.
49. Ahnazarov, S. A., Kafarov, V. V. (1985), Experiment Optimization Methods in Chemical Technology (in Russian). Moscow, Vysshajashkola, 326 s.
50. Doncov, A. A. (1978). Structuring Processes in Elastomers (in Russian). Moscow Himija. 288 p.
51. Bloh, G. A. (1972). Organic Promoters for Rubber Vulcanization (in Russian). Leningrad–Himija. 560 p.
52. Valeev, R. R. (2004). Technical Sciences Candidate's Dissertation (in Russian), Kazan, KSTU.
53. Valeev, R. R., Khakimullin, Yu. N., Byl'ev, V. A., & Liakumovich, A. G. (2002). Book of articles of the conference "Structure & Dynamics of Molecular Systems (in Russian), Jal'chik, Vol. 1, 84–87.
54. Smyslova, R. A. (1974). Liquid Thiokol Sealants (in Russian), Moscow, CNIIT Jeneftehim, 83 p.
55. Slonim, I. Ja., & Ljubimov, A. N. (1966). NMR in Polymers, Moscow 300 p.
56. Maklakov, A. I., & Pimenov, G. G. (1973), Vysokomol. Soed. (in Russian) Series A, 15(1), 107–111.
57. Nasonova, T. P., Shljahter, R. A., Novoselok, F. B., & Zevakin, I. A. (1973). Synthesis & Physical Chemistry of Polymers (in Russian), Kiev, Issue 11, 60–63.
58. Irzhak, V. I. (1975). Vysokomol. Soed. (in Russian), Series A, 17(3), 529–534.

# CHAPTER 4

# MODIFICATION OF THIOKOL SEALANTS

## CONTENTS

## 4.1   THE INFLUENCE OF EPOXY RESIN ON A MOLECULAR MOBILITY IN LIQUID THIOKOL VULCANIZATES

Increase of epoxy resin dosage is known to usually slow down vulcanization and impair a range of properties of a sealing material [1–7].

As thiokol epoxy sealants are mainly used at contact with oils and solvents, epoxy resin may extract from sealant body. It can be the main reason of unstable material's performance.

Figure 4.1 represents temperature dependences of NMR absorption line width for vulcanizates of unmodified PSO and PSO modified by E-40 epoxy resin. The analysis of obtained data indicates the increase of molecular mobility after addition of a modifier. This effect shows up in a range of lower temperatures too, while the area of abrupt narrowing of NMR line width becomes much wider, than that of unmodified vulcanizates.

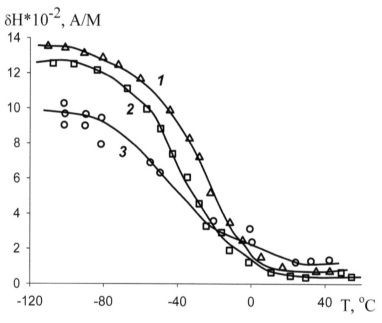

**FIGURE 4.1**   The temperature dependence of NMR adsorption line width: 1– initial PSO vulcanizate, modified by E-40 (5 mass parts); 3– PSO vulcanizate, modified by E-40 (10 mass parts).

E-40-modified PSO has better molecular mobility, and its molecular motion occurs in the range of lower temperatures. The area of NMR line abrupt narrowing is much larger, than of PSO vulcanizates. The temperature of abrupt NMR line narrowing, which characterizes initiation of developed segmental motion in polymers, reduces from −30°C for initial oligomer to −42°C U-30-MES-10 sealant. In addition, NMR absorption line for U-30-MES vulcanizates is single-component in the studied temperature range. Therefore, the temperature dependence of NMR absorption line width is lower for resin-modified sealant, than for initial PSO vulcanizate, and the interval of rapid narrowing of NMR line width shifts to a range of lower temperatures. These facts indicate that epoxy resin, added to a sealant, has a plasticizing effect, resulting in increase of molecular mobility in a system.

The analysis of spectral and temperature dependences puts forward a suggestion of weak chemical interaction of introduced epoxy resin with polysulfide oligomer. It is confirmed by obtained second moment dependence on temperature for studied polymers (Fig. 4.2).

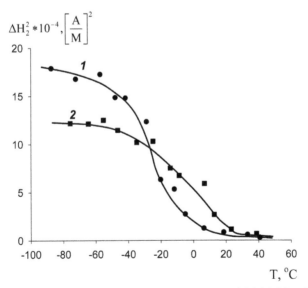

**FIGURE 4.2**   Second moment temperature dependence: 1– initial PSO vulcanizate; 2– PSO vulcanizate, modified by E-40 (10 mass parts).

The increase of second moment values at low temperatures for initial PSO vulcanizates may be caused by the following effects: PSO may vulcanize, forming defect polymer network of chains of other type, than transversal disulfide bonds. These bonds are the result of donor-coordinative interactions of oligomeric macromolecules with metal atoms of variable valence (a well-proven fact). Thus, there is the increase of intermolecular contributions to the second moment of NMR line by donor-acceptor interaction at vulcanization of polysulfide oligomers by metal dioxides. This effect can complicate the shape of NMR line for initial oligomer vulcanizates. Unlike initial oligomer vulcanizates, vulcanizates with epoxy resin do not complicate NMR line shape. It indicates the presence of polar plasticizer in a system, andTherefore, coordinative bonds become less important. In other words, plasticizer reduces the volume of in homogeneities in a system. It can affect the conversion degree of HS-groups and Therefore, change the density of forming polymer network.

Table 4.1 represents the values of NMR line second moments at low temperatures and low density of vulcanization network chains. Obtained second moment values on slope opposition to the quantity of added plasticizer. This effect can be explained by the decrease of network chain density when PSO is cured by epoxy resin. It is similar to the above described affect for plasticized compositions with PSO.

The effective chain density is the most rapidly reducing parameter, because polar additives weaken donor-acceptor interactions, occurring during formation of polymer network.

The density of chemical network chains may decrease due to decrease of polymer gel content, when a modifier is added. The experimental confirmation is provided in [8]. Observed temperature dependence of second moment for unmodified PSO also indicates, that epoxy resin is weakly chemically bound with PSO.

**TABLE 4.1** The Values of NMR Line Second Moments and Network Chain Densities of Liquid Thiokol Vulcanizates Depending on the Type of a Curing Agent and E-40 Resin Content

| Formulation | $\Delta H_2^2$, $Gs^2$ | | | $v \cdot 10^4$, mole/sm$^3$ | | | |
| | Temperature, °C | | | Effective | | Chemical | |
| | -100 | -120 | -140 | $Na_2Cr_2O_7$ | $MnO_2$ | $Na_2Cr_2O_7$ | $MnO_2$ |
| PSO (no resin) | 16.5 | 17.8 | 18.1 | 5.5 | 5.2 | 1.9 | 1.8 |
| PSO + E-40 (5 mass parts) | 14.0 | 15.1 | 15.6 | 4.2 | 4.1 | 1.8 | 1.6 |
| PSO + E-40 (10 mass parts) | 12.5 | 14.2 | 14.8 | 3.0 | 2.8 | 1.8 | 1.6 |

A plasticizing effect of epoxy resin is proved by the change of properties of sealing composition. Any mixture of PSO and epoxy oligomer is two-phase, because oligomers, their mixtures and films are turbid. We can understand the nature of a continuous phase analyzing adhesion of oligomers to various substrates. Adhesion of PSO-E-40 mixtures (with the mass ratio 20:80) to substrates is (in MPa) 0.016 for PSO and 0.051 for epoxy resin. In other words, epoxy resin is a continuous phase for the given ratio of oligomers. Therefore, 5–10 mass parts of oligoepoxide act as a plasticizer and polysulfide oligomer is a plasticizing additive for E-40 based compositions with small content of oligothiol.

It is necessary to provide the best contact of end groups to increase their conversion in the interaction of epoxy and thiol oligomeric groups. This can be only achieved, if oligomers are thermodynamically compatible. Oligomers of a various nature will otherwise segregate and end groups will be able to interact on interphase boundaries only. These phase segregations can be rather large, especially in filled compositions, which are Therefore, difficult to be made highly homogeneous. It has been determined earlier [9], that PSO-resin systems have only one area of thermodynamic compatibility, which is their mass ratio of 60:40. Meanwhile, these components are thermodynamically incompatible at temperatures below 60°C. In other words, PSO-epoxy resin system is characterized by limited mutual solubility in very narrow ratio intervals. Despite this fact, these mixtures are not inclined to delamination, because oligomers can mix finely not only via their mutual dissolution, but via mutual dispersing as well, forming a colloidal system. Temperature dependences of line widths have been used to calculate correlation times of molecular motion. Calculated $\lg \tau_c$ – $1/T$ dependence proved to be linear that is typical for indication flexible chain polymers. The values of activation energies for developed segmental motion in PSO vulcanizates and their modified products are 26.8 and 26.2 kJ/mole correspondingly.

Therefore, copolymerization and epoxy resin curing reactions are determined to be slow in PSO-E-40 compositions at the presence of vulcanizing systems, containing sodium bichromate and ethanolamine, opposite to intensive oxidation and donor-acceptor processes. The degree of system structuring is in this case similar to that of PSO-based scaling compositions. The main structuring factor in this process is PSO-epoxy resin copolymerization.

## 4.2 VULCANIZATION OF LIQUID THIOKOLS BY SODIUM BICHROMATE IN THE PRESENCE OF EPOXY RESIN

Liquid thiokol sealants, which attract a certain attention today, are compositions with a property of complete PSO and oligoepoxide chemical binding during curing process. It has been demonstrated earlier [9, 10] that vulcanizing agents, based on sodium bichromate or ethanolamine, can suit these purposes. However, the structure and molecular mobility of related vulcanizates is not studied so far. There is also no information about their curing mechanisms. Figure 4.3 demonstrates kinetic dependences for PSO-E-40 epoxy resin system, cured by sodium bichromate in the presence of ethanolamines of various basicity. Table 4.2 provides corresponding kinetic parameters for compositions with the optimal range of properties [9, 11]. Properties of vulcanizates are given in Table 4.3.

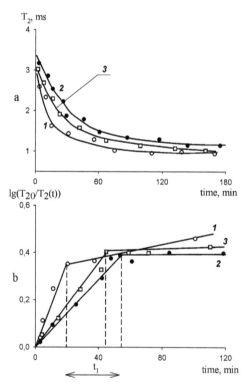

**FIGURE 4.3** Kinetic curves (a) and their semi logarithmic an amorphisms (b) for PSO + E-40 epoxy resin + sodium bichromate + ethanolamine composition; 1 – MEA; 2 – DEA; 3 – TEA.

**TABLE 4.2**  Vulcanization Kinetics Parameters for Thiokol-E-40 Sealing Compositions Depending on the Type of Ethanolamine

| Ethanolamine type | $Ki \cdot 10^3$, $min^{-1}$ | | $t_i$, min | | $T_{2\kappa}$, ms | $\Delta H_2^2$, $Gs^2$ (290 K) |
|---|---|---|---|---|---|---|
| | $K_1$ | $K_2$ | $t_1$ | $t_2$ | | |
| MEA | 49.1 | 3.41 | 17 | 102 | 0.80 | 3.5 |
| DEA | 18.7 | 1.12 | 50 | 125 | 0.90 | 3.2 |
| TEA | 21.5 | 1.26 | 45 | 115 | 0.86 | 2.5 |

Vulcanization kinetics parameters and sol content in diethanolamine (DEA) and triethanolamine (TEA)-cured compositions are similar in contrast to the compositions with monoethanolamine (MEA).

The variation of physico-mechanical properties, as well as NMR spectra final relaxation times ($T_{2final}$) and second moment values ($H_2^2$) of these vulcanizates are similar (Tables 4.2 and 4.3). This effect may be explained by the peculiarities of network structure, forming in the presence of various ethanolamines. Kinetics and structuring mechanisms have been studied for polysulfide-epoxy compositions, cured by the mixture of aqueous sodium bichromate solution with monoethanolamine, which is the most effective among above vulcanizing agents.

**TABLE 4.3**  Properties of Liquid Thiokol E-40 Vulcanizates, Depending on the Type of Ethanolamine

| Ethanolamine type | MEA | DEA | TEA | U-30–MES-10 |
|---|---|---|---|---|
| Conventional stress at 100% elongation, MPa | 2.03 | 0.91 | 0.6 | 0.86 |
| Conventional strength at rupture, MPa | 2.94 | 1.8 | 1.16 | 1.48 |
| Relative elongation, % | 220 | 300 | 540 | 440 |
| Residual elongation, % | 4 | 20 | 28 | 20 |
| Resistance to flaking from duralumin, kN/m | 3.6 | 3.8 | 2.6 | 2.0 |
| Sol content, % | 5.1 | 44.0 | 45.5 | 24.2 |

**TABLE 4.3**  *(Continued)*

| Ethanolamine type | MEA | DEA | TEA | U-30–MES-10 |
|---|---|---|---|---|
| Network chain density, $v \cdot 10^4$ mol/sm³ without E-40 | | | | |
| Without E-40 | | | | |
| effective | 0.30 | 0.16 | 0.14 | — |
| chemical | 0.16 | 0.11 | 0.11 | — |
| Network chain density, $v \cdot 10^4$ mol/sm³ with added E-40 | | | | |
| effective | 0.59 | 0.43 | 0.26 | 0.26 |
| chemical | 0.30 | 0.16 | 0.12 | 0.13 |

*30 mass parts of P-803 carbon black were added to blends, aqueous solution of $Na_2Cr_2O_7$ was used as an oxidant.

Figure 4.4 shows that the kinetic curing curve of oligomeric mixture is neither a superposition nor a multiplication of corresponding kinetic parameters, describing separate curing of oligomers. Therefore, the vulcanization mechanism of PSO + E-40 + $Na_2Cr_2O_7$ + MEA system cannot be reduced to reactions PSO + $Na_2Cr_2O_7$ and E-40 + MEA only.

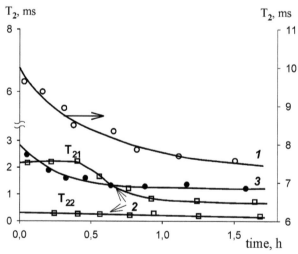

**FIGURE 4.4**  Kinetic curves for curing of PSO, E-40 epoxy resin and their mixture: 1 – PSO + $Na_2Cr_2O_7$; 2 – epoxy resin E-40 + MEA; PSO + E-40 + $Na_2Cr_2O_7$ + MEA.

If monoethanolamine is a promoter, epoxy oligomer is a retarder of PSO curing by sodium bichromate (Fig. 4.4; Table 4.4).

Epoxy resin can be cured by monoethanolamine and PSO with $Na_2Cr_2O_7$ exert a distinct influence on this process (Fig 4.5; Table 4.4).

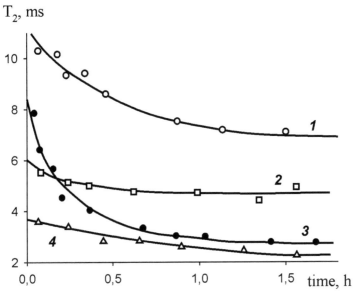

**FIGURE 4.5** PSO structuring kinetics depending on curing agent nature: 1 – $Na_2Cr_2O_7$; 2 – MEA; 3 – $Na_2Cr_2O_7$ + MEA; 4 – $Na_2Cr_2O_7$ + E-40.

The fastest process is mutual curing of oligomers (Fig 4.4; Table 4.4), although the degree of structuring is in this case similar to PSO- based compositions.

Transversal magnetization decay (TMD) of E-40-based compositions is, as a rule, two-component. According to component population data, the faster part of decay ($T_{22}$, $P_2$) is stipulated by relaxation of protons in E-40, except those in $CH_3$ groups, which contribute to the lower part with protons of other ingredients.

The decay of oligoepoxide transversal magnetization is similar: long $T_{21}$ time with population $P_1$~0.23 indicate protons of methyl groups, while short $T_{22}$ time with $P_2 = 1-P_1 = 0,77$ describes relaxation of other E-40 protons.

One must note, that P1 value of E-40 + PSO mixture at $t = t_1$ (when $T_{21}$ starts to decrease after induction period $T_{21}$ (t) = const) become equal to the population of E-40 methyl group only (0.23),that is, the rapid decay of $T_{21}$ at $t > t_1$ indicates structuring stage, which is accompanied by rapid change of intermolecular interactions in the system. $P_1$ population value after rapid decay provides easy estimation of the part of more mobile phenyl fragments. It leads to a conclusion, that less than 60% off all aromatic rings participate in network formation, while the rest of them is a potential component of a sol-fraction and (or) defective areas of a molecular network. Existence of induction period $T_{21}$(t) can be explained by initial reactions with MEA amino groups leading to self-heating of a system and distortion of observed $T_{21}$ values.

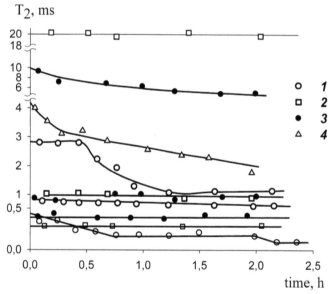

**FIGURE 4.6**  Kinetic curves of E-40 structuring at the presence of: 1 – MEA; 2 – $Na_2Cr_2O_7$; 3 – MEA + $Na_2Cr_2O_7$; 4 – MEA + PSO.

Longer component population of E-40 + $Na_2Cr_2O_7$ composition corresponds not only to fragments $C–(CH3)_3$, but assumes partitioning of water molecules to "free" (contribute to $P_1$) and "bound" (to $P_2$) as well. It can be due to reversible opening of epoxy groups in the presence of $Na^{2+}$ cations [12]:

$$\sim CH_2-CH_2 \underset{O}{\overset{}{\rightleftharpoons}} \sim \underset{OH}{CH}-CH_2-OH \quad \overset{+}{Na}, \ H_2O$$

A similar situation is observed for E-40 + MEA + $Na_2Cr_2O_7$ composition; moreover, the two-component exponential decay of magnetization in this blend obtains single-component Gauss-like shape in four days. Such a behavior of relaxation parameters in a system of almost permanent consistency can be well explained by oxidation of monoethanolamine molecules with formation of paramagnetic ions [13, 14].

**TABLE 4.4**   Kinetic Parameters of Structuring for Model Compositions

| Composition | $K_i \cdot 10^3$, min$^{-1}$ | | $t$, min | | $T_{2к}$, ms** | $T_{20}/T_{2к}$ |
| --- | --- | --- | --- | --- | --- | --- |
| | $K_1$ | $K_2$ | $t_1$ | $t_2$ | | |
| PSO $Na_2Cr_2O_7$ | 14.0 | 5.9 | 20 | 97 | 3.6 | 2.8 |
| PSO + MEA | 5.8 | 0.35 | 22 | 1400 | 3.2 | 1.9 |
| PSO + $Na_2Cr_2O_7$ + MEA | 34.3 | 1.84 | 24 | 125 | 3.3 | 2.9 |
| PSO + $Na_2Cr_2O_7$ + E-40 | 3.24 | 0.65 | 248 | 1150 | 0.83 | 4.8 |
| E-40 + MEA* | — | 39.8 | — | 80 | 0.34 | 6.6 |
| | 15.2 | 39.0 | 27 | 80 | 0.05 | 6.2 |
| E-40 + MEA+ $Na_2Cr_2O_7$* | 17.9 | 2.1 | 28 | 2190 | 0.30 | 3.5 |
| | — | — | — | — | 0.24 | 1 |
| E-40 + MEA+ PSO | 28.3 | 1.53 | 17 | 540 | 1.0 | 3.5 |
| PSO + $Na_2Cr_2O_7$ + E-40 + MEA | 49.1 | 3.41 | 17 | 102 | 0.86 | 3.5 |

* Upper values correspond to longer transversal relaxation times $T_{21}$, lower values indicate shorter times $T_{22}$. ** In one day.

Therefore, the decrease of structuring rate of E-40 + MEA blend in the presence of $Na_2Cr_2O_7$ is stipulated by inhibiting effect of alkaline oxidant's solution on E-40 epoxy groups and MEA functional centers [12, 15, 16].

Proton mobility [12] and the activity of mercaptogroups of PSO [1, 13] are known to be regulated by the content of free sulfur in the system. It makes it possible to estimate the degree of participation of PSO HS-groups in suggested processes. As Table 4.5 demonstrates, sulfur-containing groups inhibit E-40 curing by mono ethanolamine in the system with no oxidants. It can be the direct effect of blocking of E-40 + MEA reaction active centers by these groups.

**TABLE 4.5**   Kinetic Curing Parameters for PSO-basedCompositions with Varied Content of Total Sulfur*

| Composition | Total sulfur content in PSO, % | Ki·103, min-1 | $T_{2\kappa}$, ms | $T_{20}/T_{2\kappa}$ |
|---|---|---|---|---|
| PSO + $Na_2Cr_2O_7$ | 36.9 | 12.4 | 1.3 | 10.8 |
| | 37.7 | 43.2 | 1.25 | 10.4 |
| PSO + $Na_2Cr_2O_7$+ E-40 | 36.9 | 26.2 | 2.2 | 8.6 |
| | 37.7 | 45.6 | 1.74 | 8.6 |
| PSO + E-40 + MEA | 36.9 | 7.2 | 2.75 | 4.7 |
| | 37.7 | 2.7 | 2.6 | 3.0 |
| PSO + $Na_2Cr_2O_7$ + E-40 + + MEA | 36.9 | 48.4 | 1.9 | 9.0 |
| | 37.7 | 75.0 | 1.8 | 6.7 |

* At 50°C and [SH] = 1.176% (mass).

## 4.3   THE STRUCTURE OF THIOKOL COPOLYMERS WITH EPOXY RESIN. THEIR VULCANIZATION AND PROPERTIES OF RELATED SEALANTS

Liquid thiokol sealants are mainly used as protective sealants or for water-proofing of various joints. That is why they all have modifying additives, which are usually reactive and improve adhesion. One of the best-studied and widespread adhesive additives are epoxy resins [1, 2].

There are commercial sealants, based on liquid Thiokol. Their composition includes epoxy resins. They are, for example, U30 MES5 and U30 MES10 (filled P-803) sealants with E-40 epoxy Diane resin content of 6, 5 and 13 mass parts; UT-31 and UT-32 (titanium dioxide filled) 10 mass parts, correspondingly [1–5]. However, satisfactory and stable adhesion to duralumin con be observed for these blends only when they are cured with heating. One must note, that addition of epoxy resin to the formulation of sealing paste (as it's usually done in practice) reduces its pre-use storage period due to considerable increase of viscosity and related uncontrolled interaction of liquid thiokol with a resin [6]. Addition of just 6.5 mass parts of epoxy resin per 100 mass parts of thiokol reduces strength due to large quantities of uncured thiokol and epoxy resin in a system, which do not interact even at increased temperatures and act as plasticizers [7].

There are thiol-epoxy compositions, produced by thiokol and epoxy resin copolymerization in the presence of various catalysts, such as Man-

nich bases and their derivatives, combinations of organic peroxides or sodium bi chromate with amines. Professor Averco-Antonovich L. A. with
colleagues has used this principle to synthesize a series of sealants with
good strength and adhesive properties [19–24] as well as to study kinetics and mechanism of their curing [25–29]. Copolymerization is effective
in normal conditions. Considering the fact, that blends with good properties can be obtained via equimolar concentration of interacting functional groups, the ratio liquid thiokol: epoxy resin can be in the range
of 100:(20÷40), depending on the content of SH and epoxy groups. One
must note, that addition of increased quantities of thiokol resin, which are
comparable to the content of thiokol, will increase adhesion and strength
properties, but reduce elastic and low-temperature properties as well. In
addition, 100% effectiveness of copolymerization cannot be achieved in
real conditions; Therefore, such sealants always contain either excessive
or insufficient quantity of uncured oligomers that will deteriorate their resistance to solvents.

It is more preferable to add epoxy resin into liquid thiokol during synthesis [30]. An epoxy Diane resin-thiokol copolymer has been synthesized
using this approach, and corresponding formulations have been worked
out [31]. Vulcanizates of such oligomers have better strength (both adhesive and at rupture) properties, than commercial U30-MES-5 and U30-
MES-10 sealants (Table 4.6) [31–34]. Thiokol copolymer vulcanizates
have extreme dependence of basic performance characteristics on the content of epoxy component and the viability of corresponding formulations,
which are cured by commercial vulcanizing $MnO_2$-containing paste (paste
number 9), is much above the normal value (2–8 h).

As thiokol copolymer sealants have a valuable set of properties, the
study of their structure is of interest.

The composition of copolymers was studied by extraction from samples of individual oligomer components (fractions) and by determination of corresponding NMR spectra parameters. Two solvents were used,
which are substantially different in their compatibility with polysulfide
and epoxy chain links. These solvents were acetone and carbon tetrachloride (the initial concentration of solution was 20 vol. %).

TABLE 4.6 The Properties of Thioepoxy Oligomers and Related Sealants

| Label | The content of resin in a mixture of monomers, mol. % | The content of HS-groups, mass % | Viscosity, Pa × s | Conventional strength at rupture, MPa | Relative elongation at rupture, % | Residual elongation after rupture, % | Adhesion to duralumin kN/m |
|---|---|---|---|---|---|---|---|
| C–7 | 7 | 1.36 | 52.6 | 2.29 | 325 | 4 | 4.16 |
| C–13 | 13 | 2.83 | 49.8 | 2.41 | 205 | 4 | 2.87 |
| C–20 | 20 | 4.30 | 38.5 | 1.68 | 415 | 6 | 5.62 |

Therefore, copolymers with increased content of epoxy resin will form in homogeneities at the boundary between fractions when added to any of these solvents. These fractions differed considerably in viscosity and oligomeric component concentration. The last factor appeared at estimation of NMR signal levels making it possible to distinguish so called "concentrated" and "diluted" oligomeric fractions as well providing a qualitative estimation of average (the most probable) composition of low-molecular and high-molecular component in each studied solution and a copolymer in a whole [30–32].

The analysis of spectra of thiokol copolymer with 7% of ER has shown, that oligomer contains large quantities of liquid thiokol without epoxy resin in addition to copolymer structures.

The proton spectrum of a copolymer with 13% of resin, diluted in $CCl_4$, corresponds approximately to that of initial ratio of resin and polysulfide component as well as to the value of independently determined concentration of thiol groups (Fig. 4.7a; Table 4.6). 13C spectrum (Fig. 4.7b) do not contain visible signals of hydrocarbons in unopened epoxy rings (d~45 and 51.5 ppm) [35].

According to PMR spectra in acetone (Fig. 4.8), the density of phenyl fragments in the bulk of oligomer does not exceed theoretical value above 20–25%. Thus, C-13 sample is close to the a regular copolymer in the structure of macromolecular chains.

Acetone solution spectrum (Fig. 4.8c) has the signal from acidic Ph-OH proton, which does not interact in the fast proton exchange. However, the total content of C-13 component, which possibly indicates diphenylol-propan) is small.

A copolymer with 20% of resin is proved to contain no polysulfide oligomer without aromatic fragments, but its spectra demonstrate relatively intense signals, typical for methane proton of epoxy and methyl groups. A larger spread of phenyl link densities can be seen in comparison with C-13: while diluted fractions are saturated with aromatic fragments (their content exceeds theoretical value 1.5–2 times), concentrated fraction has depletion of phenyl rings (1.7 times less, than the theoretical value). It indicates the presence of unreacted ER together with ER-thiokol copolymer.

The sequence of polysulfide and epoxy block alternation in copolymers depends on oligoepoxide dosage. This sequence proved to be the most regular in the sample with 13% mass parts of epoxy resin. Nefed'ev and Mirakova also reported about these effects in their publications [36, 37].

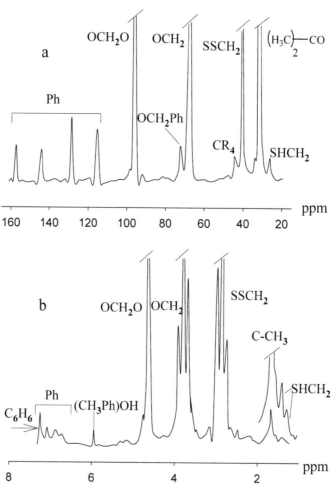

**FIGURE 4.7** PMR spectrum of C-13 copolymer: a) concentrated fraction in CCl$_4$ (+C$_6$D$_6$); b) $^{13}$C NMR in acetone.

We can suggest that copolymer structures in these oligomers form in the following processes: copolymerization via initial epoxy resin–sodium polysulfide interaction forming additional branching; interaction of a co-polymer with halogenides in the presence of sodium polysulfide, result-ing in formation of copolymer structures; ordinary copolymerization with subsequent entering of epoxy resin into reaction [1, 38–41]:

$$\sim R^1\text{-}Cl + Na_2S_n \longrightarrow \sim R^{1\text{-}}S_n\,Na^+ + \sim R\text{-}CH_2\text{-}CH\text{-}CH_2 \underset{O}{\overset{}{\diagdown}} \xrightarrow{+\,NaCl}$$

$$\xrightarrow{\phantom{NaCl}} \sim R^1\text{-}S_n\text{-}CH_2\text{-}CH\text{-}O\text{-}Na^+$$
$$\underset{\underset{R}{\overset{|}{\underset{\big\langle}{CH_2}}}}{|}$$

with a branch to NaCl.

Epoxy resin can also interact with halogenides in the following reac-tion [42–43]:

$$\sim R\text{-}CH_2\text{-}CH\text{-}CH_2 + Cl\text{-}R^1 \longrightarrow \sim R\text{-}CH_2\text{-}CH\text{-}OR^1\sim$$
$$\underset{O}{\overset{}{\diagdown}} \qquad\qquad\qquad CH_2Cl$$

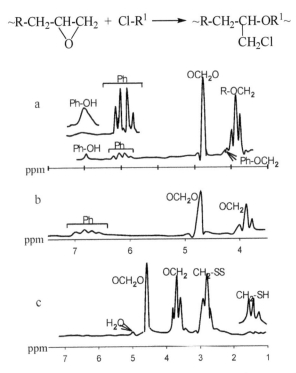

**FIGURE 4.8**  PMR spectrum of C-13 copolymer: (a) in carbon tetrachloride, lower fraction; (b) in acetone, lower fraction; (c) in acetone, upper fraction; (d) diphenylpropane in acetone.

This reaction results in additional branching. However, it occurs in alkaline media only in harsh conditions.

The synthesis of thiokol copolymer seems to involve both first and second mechanisms and lead to emerging of additional branching. Therefore, Thiokol with 13% mass of epoxy Diane resin have the most regular structure and related sealants have good set of properties. It is confirmed by NMR spectroscopy data and values of physico-chemical parameters [30,32, 42, 44].

Introduction of rigid Diane resin fragments into oligomer main chain increases the intensity of intermolecular interactions and changes its polarity. Cohesion energy density (CED) can be the parameter, characterizing polymer polarity or solubility parameter (dp), which is square root of CED [45]. Knowing this parameter, we can predict compatibility of a polymer with other polymers and plasticizers as well as its solubility in various solvents.

As Fig. 4.9 shows, a distinct maximum can be seen at $d_p$ of $(18.45\pm0.1)$ $(MJ/sm^3)^{0.5}$, for polymeric thiokol [44, 46]. Commercial liquid thiokol has lower solubility parameter value of 18.0 $(MJ/sm^3)^{0.5}$ [47].

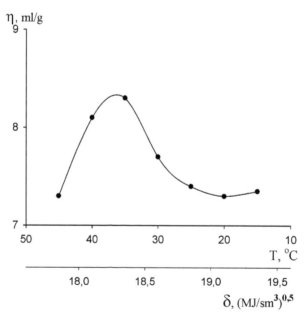

**FIGURE 4.9**  The dependence of characteristic viscosity of thiokol copolymer benzene solution on temperature and $d_p$.

Introduction of aromatic ER blocks and the increase of intermolecular interactions has been also determined to increase the viscosity of copolymer thiokol 2–2.5 times over commercial thiokol viscosity [44].

Low-temperature properties of thiokol copolymer do not differ from commercial thiokol. Its glass transition temperature is from –58 to –600°C. It seems to be due to the block structure of forming copolymer, where flexible thiokol groups are responsible for low-temperature properties.

The study of properties of sealants based on thiokol copolymers has shown, that addition of 2% mol. of trichloropropane as a branching agent (the same is for commercial thiokol), provides sealants with reduced relative elongation, while strength properties stay at high level (Table 4.7). Thus, liquid thiokol with 0,5% mol of TCP groups was used to make thiokol copolymer sealants.

Information about degree of branching of thiokol sealants is of vast interest, as it helps to predict their properties. So does the effective and chemical density of transversal bonds (DTB) with thiokol copolymer.

**TABLE 4.7**    The Influence of TCP Content in Thiokol on Properties of Sealants

| TCP content, mol. % | Conventional strength at rupture, MPa | Relative elongation at rupture, % | Relative residual elongation after rupture, % | Adhesion to duralumin kN/m |
|---|---|---|---|---|
| 0,5 | 2.93 | 425 | 6 | 4.6 |
| 1,0 | 2.80 | 350 | 4 | 4.2 |
| 1,5 | 3.12 | 225 | 0 | 4.3 |
| 2,0 | 2.97 | 160 | 0 | 3.9 |

DTB has been estimated for commercial sealants filled with carbon black and E-40 epoxy Diane resin content of 6.5 and 13 mass parts (U30-MES-5 and U30 MES-10 correspondingly) and for thiokol copolymer with 13% mol. of E-40 epoxy resin. Epoxy resin was being introduced during the synthesis of a sealant. The chemical density of transversal bonds ($n_{chem.}$) was estimated after the degree of swelling of sealants in toluene reached equilibrium. DTB estimation results are represented in Table 4.8 [46].

As this table demonstrates, when the content of epoxy resin in commercial sealants increases, the quantity of chemical bonds reduces promptly. The theoretical value of density of chemical bonds in vulcanizates with 2 mol. % TCP is $2.3 \times 10^{-4}$ mol/sm$^3$ [49]. However, the experimental nchem. value is considerably below, than the theoretical one. This is because of inhibiting effect of epoxy resin on thiokol curing and resulting increase of uncured thiokol and epoxy resin quantities according to sol-gel analysis data as well as substantial deterioration of strength, hardness and performance characteristics [7].

The theoretical value of chemical bond density is $0.58 \times 10^{-4}$ mol/sm$^3$ for sealant with 0.5% mol. of TCP, taking the content of epoxy resin in a copolymer into consideration. As (Table 4.8) shows, the density of chemical bonds in thiokol copolymer sealant exceeds theoretical value over 20%. It can be explained by additional branching during thiokol synthesis by above reaction.

**TABLE 4.8**   The Influence of Sealant's Composition on DTB Value and Its Properties

| Component* | Curing regime | Density of transversal bonds n•104 mol/sm3 | | | Viability, hours | Conventional strength at rupture, MPa | Relative elonga-tion at rupture, % | Relative residual elonga-tion after rupture, % | Resistance to flaking from D-16 duralumin kN/m |
|---|---|---|---|---|---|---|---|---|---|
| | | effective | chemical | physical | | | | | |
| 1 | 24 h at 70°C | 1.34 | 0.57 | 0.77 | 5 | 2.11 | 440 | 8 | 5.34 |
| 2 | | 0.40 | 0.08 | 0.32 | 4 | 1.72 | 450 | 10 | 3.8 |
| 3 | | 3.36 | 0.73 | 2.63 | 22 | 2.57 | 400 | 6 | 5.0 |
| 1 | 7 days at 20°C | 1.34 | 0.56 | 0.78 | - | 2.28 | 410 | 7 | 1.5 |
| 2 | | 0.47 | 0.12 | 0.35 | - | 1.85 | 470 | 10 | 1.1 |
| 3 | | 3.30 | 0.74 | 2.56 | - | 2.68 | 360 | 4 | 6,3 |

*1–U30MES-5 sealant, 2–U30MES-10 sealant, 3–thiokol copolymer sealant.

On must also note a considerable increase of physical bonding in thiokol copolymer sealants because of epoxy polar fragments б included into oligomer's main chain (Table 4.8).

The properties of U30 MES-10 and UT-32 sealants based on thiokol–epoxy Diane resin (13% mol.) copolymer have been compared with the properties of commercial sealants of U30 MES-10 and UT-32 grades, containing 13 and 10 mass parts of epoxy Diane resin [48]. Table 4.9 shows properties of thiokols used in production of above sealants.

**TABLE 4.9** The Properties of Thiokol Copolymer and Commercial Thiokol's

| Thiokol type | The content of SH-groups, weight % | The content of a branching agent, mol. % | Viscosity, Pa•s |
|---|---|---|---|
| Copolymer | 3.30 | 0.5 | 37.2 |
| 1-NT | 3.05 | 2.0 | 18.5 |
| 2-NT | 2.33 | 2.0 | 34.8 |

As Fig 4.10 shows, the rate of sealant vulcanization, estimated via Shore hardness a growth, is higher for Thiokol copolymer. These sealants reach maximum hardness in 4 days already, while this parameter for commercial sealants is 7 days. Higher hardness values for thiokol copolymer sealants (10–15 units) can be explained by the presence of plasticizing unreacted epoxy resin even in fully cured commercial sealants.

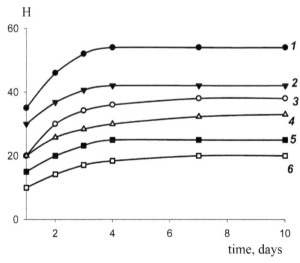

**FIGURE 4.10** Shore hardness variation kinetics for thiokol copolymer sealants (1, 3, 5) and commercial thiokols (2,4,6). 1,2–carbon black; 3,4–titanium dioxide; 5,6-chalk.

Data from Table 4.10 demonstrate, that thiokol copolymer sealants overcome commercial serial products in conventional strength at rupture (by 20–30% in average) and in adhesion to duralumin after non-heating vulcanization.

Thiokol sealants are used in many cases at increased temperatures; Therefore, the analysis of variation of their properties in a temperature range of 20–100°C seems to be important. As we can see from Table 4.10, thiokol copolymer sealants keep their set properties good at all studied temperatures. Properties vary independently on thiokol type. However, sealants with different fillers demonstrate individual behavior.

**TABLE 4.10**   The Properties of Thiokol Copolymer and Industrial Thiokol Sealants

| Parameters | Testing temperature, °C | U30 MES10 | | UT-32 | | AM-05 | |
|---|---|---|---|---|---|---|---|
| | | copolymer | commercial | copolymer | commercial | copolymer | commercial |
| t, hours/minutes | | 3h 10 m | 2h 00 m | 3h 50 m | 2h 10 m | 5h 40 m | 3h 10h |
| s, MPa | | 2.3 | 1.9 | 2.3 | 1.7 | 0.4 | 0.2 |
| e, % | 23 | 240 | 290 | 220 | 180 | 360 | 400 |
| A, kN/m | | 3.4 | 3.5 | 3.4 | 1.5 | 3.1 | 1.2 |
| s, MPa | 50 | 2.2 | 1.8 | 1.6 | 1.2 | 0.24 | 0.11 |
| e, % | | 100 | 80 | 110 | 90 | 130 | 160 |
| s, MPa | 70 | 1.6 | 1.2 | 1.4 | 1.0 | 0.2 | 0.1 |
| e, % | | 90 | 65 | 50 | 40 | 115 | 140 |
| s, MPa | 100 | 1.0 | 0.6 | 1.1 | 0.9 | 0.19 | 0.1 |
| e, % | | 88 | 60 | 50 | 40 | 105 | 120 |

*Accelerated curing regime (+70°C, 24 h).

Properties of formulations with titanium dioxide deteriorate with increase of testing temperature and stabilize at 70°C. Formulations with carbon black keep their strength at temperatures up to 50°C that can be due to the presence of dense network of physical interactions, which weakens substantially after further increase of temperature. Sealants with different fillers have equal strength behavior if total lack of physical interactions is observed (at 100°C).

Compositions filled with carbon black overcome ones with titanium dioxide in their relative elongation by approximately 1.5 times at 100°C, that is due to increased content of oligomer in compositions with carbon black.

Outdoor aging of sealants has been studied at temperatures of 70, 100 and 125°C (Fig. 4.11). Thiokol copolymer sealants have been determined to be less inclined to such aging. Thermo-oxidative aging is more intensive for sealants filled with titanium dioxide. It should be noted, that commercial sealants demonstrate deeper deterioration of properties in these conditions, probably due to epoxy resin structuring during aging.

Data on the behavior of sealants in solvents and water are highly informative (Table 4.11). As we can see, Thiokol copolymer sealants swell much less in water (2.5–4 times) and toluene (1.5 times).

TABLE 4.11    Kinetics of Sealant Swelling in Water and Toluene, %

| Sealant | Water | | | | | Toluene | | | | |
|---|---|---|---|---|---|---|---|---|---|---|
| | 1 day | 3 days | 7 days | 10 days | 16 days | 1 day | 3 days | 7 days | 10 days | 16 days |
| U-30 MES-10 copolymer | 1.2 | 1.8 | 2.6 | 2.9 | 3.5 | 33.8 | 40.1 | 40.4 | 41.1 | 43.0 |
| U-30 MES-10 commercial | 3.8 | 6.6 | 9.6 | 12.0 | 16.2 | 59.2 | 70.9 | 71.3 | 72.2 | 73.7 |
| UT-32 copolymer | 1.3 | 1.9 | 2.7 | 3.2 | 4.7 | 37.5 | 38.8 | 39.2 | 39.8 | 42.8 |
| UT-32 commercial | 2.8 | 4.9 | 8.1 | 10.3 | 13.1 | 64.3 | 67.2 | 70.9 | 72.3 | 74.9 |
| AM-05 copolymer | 1.6 | 2.7 | 3.8 | 4.4 | 5.2 | 17.3 | 21.8 | 22.9 | 23.7 | 24.1 |
| AM-05 commercial | 2.5 | 4.3 | 7.5 | 9.2 | 11.4 | 29.4 | 34.1 | 36.7 | 38.1 | 38.6 |

Swelling behavior and value are independent on the nature of filler. It is due to higher network chain density in Thiokol copolymer sealants, than in serial commercial formulations [48].

Thiokol copolymer sealants keep higher performance after aging in water. This most sensitive parameter is in this case adhesion to duralumin. Its value stays almost unchanged for Thiokol copolymer sealants, while commercial serial formulations have this parameter reduced considerably and delamination type changes from cohesive to adhesive. Strength and relative elongation increase monotonically during first 3–5 days. Further aging in water does not almost deteriorate strength properties. Sealants with carbon black lose their properties in a less degree, than sealants with titanium dioxide when aged in water.

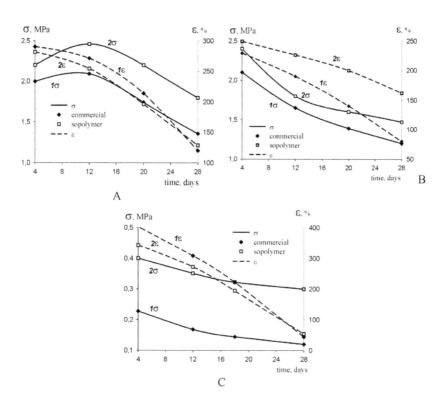

**FIGURE 4.11** The variation of conventional strength at rupture (–), relative elongation (—), at air aging (125°C) of U-30 MES10 (a), UT32 (b) and AM-05 (c) sealants based on commercial (1) and copolymer (2) thiokols.

Thiokol-epoxy resin copolymers seem to be rather prospective for making of sealants for sealing interpanel joints in house building. Thiokol copolymer sealants have been compared with AM-05-grade sealant, based on commercial TSD thiokol with added 5 mass parts of E-40 epoxy resin.

Thiokol copolymer sealants have been determined to have better strength, adhesion to concrete and hardness. More effective vulcanization results in formation of sealants with higher density of both physical and chemical transversal bonds and less swelling in water and toluene, respectively (Table 4.12).

One should note that sealing pastes with Thiokol copolymer are more stable at storage. The predictable guaranteed storage period is 12 months for such pastes, while this value for commercial serial U30 MES5, U30 MES-10 and AM-05 sealants is 3 months. Sealants for joints are usually applied into joint with syringe; Therefore, stable viscosity during all the storage period is a very important parameter. It is possible if Thiokol copolymer is used. If epoxy resin is introduced separately, viscosity increases considerably already after one month of storage, that deteriorates the quality of a mastic seam.

TABLE 4.12   Properties of Thiokol Copolymer and Commercial Sealants

| Parameters | AM-05, copolymer | AM-05, commercial |
|---|---|---|
| Viability, hours | 15 | 6 |
| Strength, MPa | 0.40 | 0.23 |
| Relative elongation, (on seams) % | 180 | 240 |
| Adhesion to concrete, MPa | 0.95 | 0.65 |
| Shore hardness, units | 25 | 15 |
| Water absorption after 40 days, % | 7 | 13 |
| Welling in toluene after 20 days, % | 24 | 38 |
| Density of transversal bonds, $\times 10^4$ | | |
| Effective | 1.33 | 0.83 |
| Chemical | 0.6 | 0.27 |

Therefore, we have studied the composition of Thiokol copolymers with various content of epoxy Diane resin. Such Thiokol's have been determined to mainly consist of a copolymer. It has the most regular structure at the content of epoxy resin in a copolymer of 13% mol.

Properties of Thiokol copolymer have been studied. Epoxy resin additives have been determined to increase solubility parameter of a copolymer. The density of transversal chains in a sealant has been estimated as well as contributions of chemical and physical interactions. We have proposed the mechanism of epoxy resin interaction with sodium polysulfide and di-b-chloroethylformal during polycondensation and demonstrated its ability to form additional branching as TCP does.

Properties and behavior of sealants have been studies in various aging conditions. Thiokol copolymer sealants have been shown to overcome commercial serial grades in by physico-mechanical and adhesive properties, curing rate and resistance various types of aging. These copolymers can be used for production of sealants for various branches of mechanical engineering and construction industry.

## 4.4   MODIFICATION OF THIOKOL SEALANTS BY UNSATURATED POLYESTERS

Glass packets find various applications in construction industry today. Sealants for glass packet outlines should possess gas impermeability, good physico-mechanical properties as well as high adhesion to glass and duralumin. However, Thiokol two-component sealant 51-UT-48, used for this purpose, does not have a property of high and stable adhesion, especially to glass.

A prospective alternative is modification of liquid Thiokol's by unsaturated monomer or oligomer compounds [1, 50], which does not only increase adhesive properties, but also can even form copolymer sealants with new range of properties.

The analysis of processing conditions and the level of properties, required from sealants for glass packets revealed requirements of adhesive additives: they are supposed to have functional groups interacting with Thiokol during curing and simultaneously form adhesion bonds with glass and duralumin as well remain inert to Thiokol SH-groups when stored before application. Combination of these controversial requirements in one compound was successful thanks to usage of unsaturated polyesters with carboxyl end groups.

Indeed, abietic acid, its derivatives and oligoesteracrylates (OEA) considerably increase its adhesion to metal and glass [51–55]. Unsaturated compounds act as temporary plasticizers and actively participate in vulcanization [56]. Reactive compounds are introduced into sealing paste with Thiokol. It increases paste's viscosity with time and leads to considerable shortening of its pre-use storage time. Therefore, we have studied an opportunity of application of unsaturated polyesters (USPE), which do not interact with Thiokol's end SH-groups in conditions of combined storage and are able to participate actively in curing of Thiokol. At the same time, the efficiency of unsaturated polyesters of various structures has been estimated for reactions with liquid Thiokol depending on the composition and nature (activity) of constituent double bonds.

Commercial first grade liquid Thiokol has been used. The content of SH- groups was 2.95% mass and the viscosity value was 15.5 Pa·s. Hydrophobic chalk was used as a filler in the quantity of 80 mass parts. Liquid Thiokol was cured by manganese dioxide in the form of a curing paste. The mass ratio of sealing and curing pastes was in this case 100:10. Other components were pine colophony, oligoesteracrylates (TGM-3-tri(oxyethylene)-α,ω-dimethylacrylate, MGPh-9-α,ω-dimethylacrylate-(bis-ethyleneglycolphthalate) and USPE of various molecular weight and composition and with carboxylic end groups. 0.3–10 mass parts of USPE per 100 mass parts of thiokol were added to a sealing paste.

As Table 4.13 shows, viability of sealants increases with addition of unsaturated compounds, such as colophony, which mainly consists of a biotic acid and USPE.

Application of all these compounds results in the increase of strength properties, adhesion to glass and duralumin, decrease of relative elongation of sealants. It indicates that studied additives participate in liquid Thiokol curing reactions thanks to their double bonds. Curing has been determined to have radical mechanism [56], and the process is initiated by inorganic peroxide (manganese dioxide) and tertiary amine (diphenylguanidine). One must note, that colophony and oligoesteracrylates have insufficient efficiency that can be due to low activity of double bounds to end SH-group at low-temperature curing (20–300°C). Adhesive, deformation and strength properties are deeper influenced by USPE [57–65]. When acid number and corresponding content of carboxyl groups increase, viability enlarges (Table 4.13).

**TABLE 4.13**   Properties of Thiokol Sealants Modified by Unsaturated Compounds

| Sample | Quantity, mass parts | Viability, hours | Conventional strength at rupture, MPa | Relative elongation % | Adhesion to glass, MPa |
|---|---|---|---|---|---|
| Reference | 0.0 | 050 | 0.93 | 285 | 0.51 |
| | 1.0 | 120 | 1.03 | 270 | 0.65 |
| | 2.0 | 155 | 1.30 | 245 | 0.78 |
| Colophony | 3.0 | 220 | 1.37 | 220 | 0.84 |
| | 5.0 | 345 | 1.32 | 205 | 0.9 |
| TGM-3 | 5.0 | 200 | 0,81 | 350 | 0.77 |
| MGPh-9 | 5.0 | 245 | 0.97 | 340 | 0.79 |
| USPE, acid number:29.5 | 2.0 | 245 | 1.39 | 210 | 1.05 |
| 50.0 | 2.0 | 310 | 1.44 | 165 | 1.11 |
| 80.0 | 2.0 | 355 | 1.46 | 120 | 1.24 |
| 120.0 | 2.0 | 520 | 1.58 | 75 | 1.36 |

When the content of resin in sealant increases, as well as resin's acid number grows, sealant becomes more adhesive to duralumin and glass and strong, it relative elongation reduces [61]. It is stipulated in the first case by the increase of concentration of active unsaturated bonds and in the second case by accelerated diffusion of resin and its more active interaction involving double bonds, increase of content of carboxyl groups able to form chemical bonds on sealant-duralumin and sealant–glass boundaries.

Further research has been carried out with unsaturated polyester resin (0.5–10 mass parts) with the acid number of 29.8 mg KOH/g.

It has been proved (Fig. 4.12) that, maximum change of all studied properties can be observed if such resin is added in the quantity of 2.0–3.0 mass parts. Changes in viability and adhesion are similar; these parameters gradually increase when resin content in a sealant exceeds 2.0 mass parts.

Conventional strength at elongation has extreme and its maximum is at the same additive of resin. Resin content above 2 mass parts leads to insignificant change of strength. Relative elongation at rupture increases even with small additives of resin. The described behavior of properties seems to be due to active interaction of resin double bonds with Thiokol SH-groups during curing and formation of 3-dimensional network. Indeed, the chemical density of network chains ($n_{chem.}$) grows with polyester concentration (Fig. 4.13). When USPE content is above 1 mass part, $n_{chem}$. exceeds its calculated value of $2.3 \times 10^{-4}$ mol/sm3 corresponding to TCP concentration in Thiokol of 2% mole.

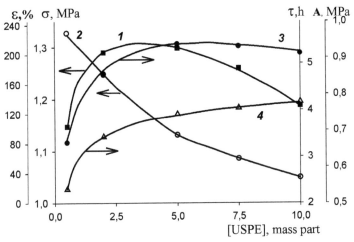

**FIGURE 4.12** The influence of USPE content on conventional strength at rupture (1), relative elongation at rupture (2), viability (3) and adhesion to glass (4).

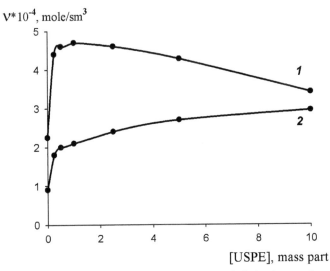

**FIGURE 4.13** The dependence of density of transversal chains in a sealant on USPE content: 1 – effective; 2 – chemical.

USPE can be suggested to cause additional sealant structuring, because its double bond functionality exceeds 3. In the same time, the effective density of transversal bonds $n_{eff}$ increases. The change of $n_{eff}$ with USPE content has extreme behavior, because physical interactions between polar Thiokol groups and a filler are opposed by spatial obstacle in a form of bulky aromatic USPE structures added in the number 1–2 mass parts. It's not only USPE concentration that influences its efficiency in sealants, but the nature of double bond as well, which is specified by introduced co-monomers with double bonds [64]. Polyesters with maleic (MA), endic (EA) and phthalic anhydrides nave been synthesized both individually and in various combinations to estimate the influence of polyester's double bond nature properties of sealants. Synthesized USPE have been tested by $^1H$ and $^{13}C$ NMR–spectroscopy. $^1H$ spectra of polyesters of various structures are given in (Fig. 4.14). They have three distinguishable areas: the area of olefin protons of endic and maleic anhydride residues (5.9–6.9 ppm) the area of glycol residues (3.0–5.0 ppm) and the A change of polyester composition effects their structure and specifies the content of norbornene (endic anhydride) and maleic-fumaric groups in polyester. The ratio of saturated and unsaturated acidic reactive components determines a potential activity of polyester cross-linking process during their further usage, because participation of USPE in0020PSO curing mainly depend on the content of double bonds in polyesters of various structure.

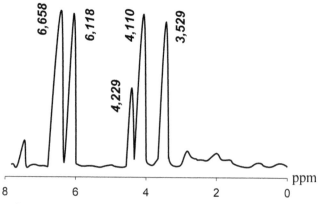

**FIGURE 4.14** $^1H$ NMR spectra of USPE with the following composition: EA: MA=0.9:0.1.

Therefore, we were first of all interested in the content of double bonds in polyesters. Although proton spectra are easy enough for determination of anhydride-glycol group ratio, these spectra are less informative than $^{13}$C NMR ones due to overlapping of signals. Typical $^{13}$C NMR spectra of studied polyesters are given in Fig. 4.15. Three areas of signals can also be found in these spectra: 1–the area of signals from $CH_2$ and CH groups in endic anhydride residues (d=40÷55 ppm); 2–signals from groups of diethyleneglycole and ethylene glycol residues as well as $CH_2OH$, and $CH_2OCH_2OC(O)$ (d=71,86 ppm); 3–the area of signals from CH=CH in endic and maleic anhydride residues (d=125÷140 ppm). The most informative $^{13}$C spectra areas for our purposes are areas of olefinic carbons in endic or maleic anhydrides and oxyaliphatic groups of glycols. The analysis if $^{1}$H and $^{13}$C spectra shows, that the content of active double bonds in polymers decreases in symbiosis with the change of polyester composition, if EA content in polyester is reduced from 0.9 to 0.1 molar parts. A reliable estimation of polyester composition using the method of quantitative calculation of integral intensities by $^{13}$C NMR spectra is only possible if signal areas of group of the same type are compared (C=C area), as it eliminates mistakes arisingdue to difference in relaxation times and overhauser nuclear effect [63, 64]. The reduction of EA content in polyester from 0.9 to 0.1 molar parts results in 30% reduction of CH=CH group content in a polyester, that will further influence curing kinetics and properties of modified thiokol sealants, when these polyesters are used. One should also note the presence of free EA signal in polyester spectra from CO (O) group ~169–172 molar parts. The content of EA is small, indicating its almost complete conversion during synthesis.

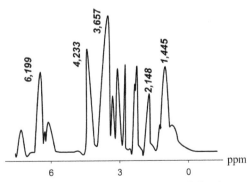

**FIGURE 4.15** $^{13}$C NMR spectra of USPE with the following composition EA: MA=0.1:0.9.

The change of MA to FA group in polyester reduces the quantity of double bonds in PE because of aromatic chain links, though the general structure of this PE stays similar to the corresponding unmodified PE structure. Therefore, the properties of source polyesters and their activity in reactions with PSO SH-groups will depend not only on the nature of individual components, forming polyester chain, but on their distribution along PE chain as well. So, PSO composition can deviate from the composition of blend of initial components die to different content of fumarate or maleinate double bonds or the absence of these groups in PE of high molecular weights. It will certainly effect their activity in various curable oligomeric compositions able to interact with PE.

Figure 4.16 represents kinetic curves of curing a blend with polyesters of various composition. The kinetics of a curing process was measured using pulse NMR by the nuclear spin-spin relaxation time $T_2$. The analysis of kinetic curves indicates, that USPE blends with high content of endic anhydride can be cured faster and, in addition, change of MA to FA retards curing process as well.

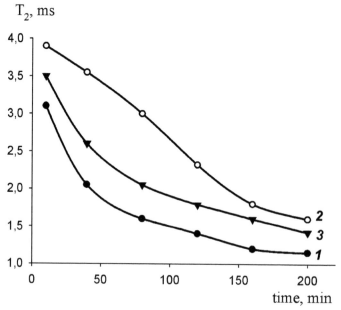

**FIGURE 4.16**    Kinetic curing curves of compositions of various type: 1–EA:MA=0.9:0.1; 2–EA:MA=0.1:0.9; 3–EA:F=0.9:0.1.

The final spin-spin relaxation time $T_2$ is in good correlation with effective chain density ($n_{eff}$) of a network forming in cured sealing compositions: the increase of a network chain density should lead to shorter spin-spin relaxation time $T_2$. Thus, it is possible to calculate the effective network chain density via found $T_2$ time values.

Table 4.14 represents found correlations of final $T_2$ times with the effective network chain densities of compositions of various types.

Such an influence of USPE structure on curing kinetics correlates well the structure of latter compounds as well as their quantity and activity of double bonds determined in high resolution NMR study.

**TABLE 4.14** The Dependence of Effective Network Chain Density on Spin-Spin Relaxation Time $T_2$

| Polyester | $T_2$, ms | $v_{eff}*10^4$, mole/sm$^3$ |
|-----------|-----------|------------------------------|
| EA:MA = 0.9:0.1 | 1.20 | 2.3 |
| EA:MA = 0.1:0.9 | 1.55 | 1.6 |
| EA:FA = 0.9:0.1 | 1.40 | 2.1 |

Application of USPE with EA favors the increase of transversal chain density in sealants [64]. The decrease of EA content leads to 30% decrease of a network chain density. It confirms the fact, that MA double bonds are insufficiently effective in relatively mild curing conditions, while EA double bonds complete (quantitatively) interaction with Thiokol's end SH-groups. EA is the most preferable among studied anhydrides: EA→MA→FA, because its application results in higher network chain density and better physico-chemical parameters of cured blends. Obtained data are confirmed by properties of USPE-modified thiokol sealants of various compositions [65]. As Fig. 4.17 shows, introduction of double bonds into PE macromolecule with MA co-monomer leads to considerable increase of sealant's strength and adhesion to glass as well as to decrease of relative elongation. EA leads to stronger change of properties, than MA (Fig. 4.18).

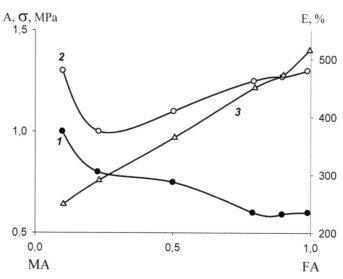

**FIGURE 4.17** The dependence of sealant's adhesion–1, conventional strength at rupture–2 and relative elongation–3 on USPE composition.

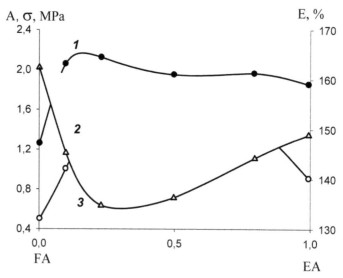

**FIGURE 4.18** The dependence of sealant's adhesion–1, conventional strength at rupture–2 and relative elongation–3 on USPE composition.

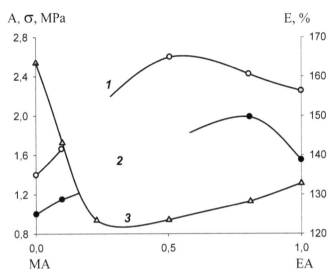

**FIGURE 4.19** The dependence of sealant's adhesion–1, conventional strength at rupture–2 and relative elongation–3 on USPE composition.

The nature of double bonds exerts a visible effect if polyesters with both maleic and endic anhydrides are added to a sealant. If as little as 10% of MA are changed to EA, strength and adhesive properties increase. However, the relative elongation changes to a little degree (Fig 4.9).

We think, that the effectiveness of USPE as a modifier for thiokol sealants is first off all stipulated by the presence of double bond electronic density. For example, oligoesteracrylate's end double bond belongs to isopropenyl group and, in addition, conjugated with carboxylic group. It reduces its activity in reactions of both interaction with SH-groups and vulcanization.

MA in polyester macromolecules opens during synthesis forming fragments of maleic and fumaric acid fragments. Depending on synthesis conditions, there are some double bonds with cis- and trans-configuration in a molecule, which, despite conjugation, participate actively in above reactions.

Finally, unsubstituted double bond introduced into PE with MA molecule, is situated in five-atom ring and is not conjugated with any group having active electrons. In addition, ring enters into some conjugation, which makes double bond even more active endic anhydride molecule

When USPE with carboxylic end groups is added to a sealing paste, one may expect the increase of stability at pre-use storage thanks to inhibited oxidation processes.

Seven batches of a sealing paste have been made to estimate properties and stability at storage of sealing paste with unsaturated polyester resin. The quantity of added resin was 0.3 mass parts per 100 mass parts of Thiokol. As addition of resin inhibits vulcanization, manganese dioxide was added to a curing paste and diphenylguanidine promoter was added twice. Results are given in Table 4.15 [61].

The tabledemonstrates, that modified sealant substantially overcomes standard sealant 51-UT-48 in strength and adhesive properties. The pattern of variation of penetration indicates little interaction of resin with Thiokol at storage. A predictable storage time may be six months for designed formulations.

Therefore, USPE composition has been analyzed using $^1$H and $^{13}$C NMR methods and it has been determined to participate in curing of Thiokol sealants as a structuring agent and its activity depends on the nature of double bonds. Addition of 1–3 mass parts of USPE to thiokol sealants increases their adhesion to glass and duralumin and conventional strength at rupture. Addition of USPE with EA provides sealants with better range of properties.

**TABLE 4.15**   Changing of USPE-Modified Sealant Properties at Storage

| Parameter | 51–UT-48 | Modified sealant's storage time, month | | | |
| --- | --- | --- | --- | --- | --- |
| | | 0 | 1 | 3 | 6 |
| Viability, minutes | 60–480 | 160 | 165 | 175 | 195 |
| Conventional strength at tension, MPa | 1,0 | 1,46 | 1,79 | 2,06 | 1,91 |
| Relative elongation at rupture, % | 160 | 240 | 217 | 205 | 195 |
| Tear strength of sealant applied on glass or duralumin, MPa* | 0,7 | 1,12 | 0,92 | 0,98 | 0,89 |
| Penetration, mm$^{-1}$ | 120 | 326 | 310 | 295 | 282 |

*Cohesive type of tear.

## 4.5  COMPOSITION, CURING AND PROPERTIES OF THIOURETHANE SEALANTS

Single and two-component urethane sealants are very popular in construction industry today. Urethane prepolymers with NCO end groups on the basis of polyoxipropyleneglycols (laprols) (SKUPPL-4503, SKUPPL-5003) are widespread components for production of two-component formulations because of easily available feedstock sources and cheapness of these PEs. These prepolymers are large-scale products of "Tenax" (Latvia ), "Germoplast,""SKIM" (Moscow) and are used in construction as roofing formulations as well as for sealing of interpanel joints. It was interesting to work out thiourethane compositions using such reactive oligomers. They would have good deformation properties and adhesion to various construction materials. The dependence of curing kinetics and properties of formulations on the ratio and nature of oligomers has been studied to make such sealants with optimal properties. Basic characteristics of corresponding oligomers are given in Table 4.16.

**TABLE 4.16**    The Properties of Oligomers Used for Synthesis of Thiourethane Sealants

| Oligomer grade | Polyester | The content of end groups, % | h(20°C), Pa*s | dp, (MJ/ m³)$^{1/2}$ |
|---|---|---|---|---|
| SKUPFL – 100 | polytetramethyleneglycol | 5.85 (NCO) | 11.5 | 18.3 |
| SKUPPL – 5003 | Laprol – 5003 | 2.34 (NCO) | 8.9 | 16.9 |
| SKUPPL – 4503 | Laprol – 4503 | 2.65 (NCO) | 7.7 | 16.9 |
| PIC | - | 30 (NCO) | 5.2 | 21.2 |
| NVB – 2 thiokol | - | 3.52 (SH) | 10.2 | 18.0 |
| TPM – 2 | Laprol – 4503 | 2.1 (SH); 1.08(OH) | 3.0 | 16.9 |

Compatibility of components was considered when selecting, as only good compatibility of these oligomers will result in obtaining of compositions with good range of properties. A reliable characteristics of compatibility of polymers could be solubility parameter (dp) [66, 67]. As Table 4.16 shows, all isocyanate-containing components excluding polyisocyanate (PIC) are supposed to be compatible with PSOs. Among all isocyanate-containing components under study, SKUPFL-100 has the solubility parameter closest to the that of liquid thiokol, while laprol prepolymers are fully compatible with TPM-2 polymer. Solubility parameter of PIC is somewhat higher, than of thiokol and TPM-2 polymer. But as the content of PIC in a formulation is ten times, than of PSO, the difference in polarity is nor so important.

### 4.5.1 THE INFLUENCE OF THE NATURE AND THE RATIO OF COMPONENTS ON CURING KINETICS

The curing kinetics of thiourethane formulations depends on many factors, mainly on the nature of oligomers, their ratio and compatibility, the nature and content of catalyst and a filler [19, 46, 68–70].

2,4,6-tris (dimethylaminomethyl) phenol (OM–3) was used as a catalyst. As curing was catalyzed by a tertiary amine, formation of thiourethanes was a primary reaction in these systems [71, 72].

Curing kinetics of thiourethane compositions has also been studied by NMR spectroscopy [73, 74]. Figure 4.20 represents $T_2$ variation with time for formulation with thiokol and various ratio of components. Formulations with equimolar ratio of SH and NCO groups are seen, to have the maximum rate at the initial period of curing (See Figs. 4.20 A, B, C). It may be because the excess of any of components acts as a plasticizer for a composition in the initial period and inhibits curing. When PIC is used, the initial period is much shorter and the curing rate of compositions with PIC excess (NCO:SH–1,5:1) is higher, than of composition with equimolar ratio of functional groups in three hours already. It is due to a low molecular weight and high functionality of PIC (f = 2.5 , 3.0), andTherefore, viscosity can substantially grow in the initial period because of occurring structuring and formation of network structures.

The curing rate of compositions with well compatible components (Thiokol + SKUPFL-100) is substantially higher, than of poorly compatible formulations (Thiokol + SKUPPL-5003).

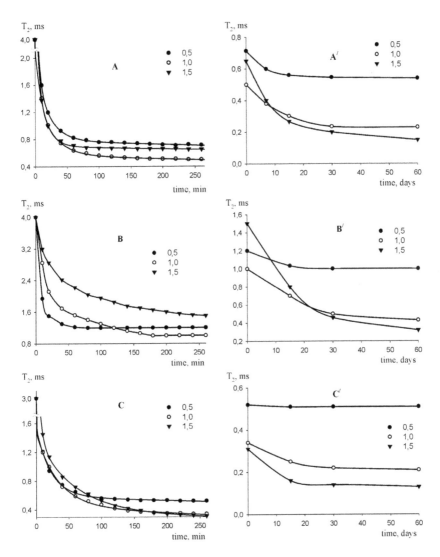

**FIGURE 4.20**   The dependence of $T_2$ on curing time (A,B,C–the initial stage; A', B', C'–the final stage) of thiourethane sealants based on liquid thiokol (A, A'–with SKUPFL-100; B, B'–with SKUPPL-5003; C, C'–with PIC) at the ratio of NCO: SH : ● – 0.5; ○ – 1; ▼ – 1.5.

The second stage of curing (from the moment of gel formation to complete curing) differs both in duration and behavior for formulations with various isocyanate-containing components and NCO:SH ratios (See

Fig. 4.20 A–C). When PIC is changed to a prepolymer, curing rated decreases due to slower formation of secondary bonds, than in reaction of thiourethane formation [74–76]. Figure 4.20 demonstrates secondary reactions occurring at the extent of NCO: prepolymer, unreacted at the curing first stage, enters with time in a reaction forming network structures and thus making a sealant with increased degree of curing. Thioallphanate transversal bonds form in the following reaction:

$$\sim R\text{-NCO} + \text{HS-}R'\sim \longrightarrow \sim R\text{-N(H)-C(O)-S-}R'\sim$$

$$2\sim R\text{-N(H)-C(O)-S-}R'\sim + \text{NCO-}R''\text{-OCN} \longrightarrow$$

$$
\sim R-N-\overset{\overset{\displaystyle (O)}{|}}{C}-S-R'\sim
$$

$$(O)C-\overset{\overset{\displaystyle H}{|}}{N}-R''-\overset{\overset{\displaystyle H}{|}}{N}-C(O)$$

$$\sim R-N-\overset{\overset{\displaystyle (O)}{|}}{C}-S-R'\sim$$

The curing kinetics of formulations with TPM-2 polymer is both similar to and different from formulations with thiokol (Fig. 4.21). The presence of both hydroxyl and SH-groups in TPM-2 polymer seems to result in their mutual activation [74] and higher reactivity of such formulations. Thus, they achieve the state of complete curing faster, than formulations with thiokol. As TPM-2 polymer contains one SH-group per one OH-group, the required equimolar ratio relative to isocyanate is 2NCO:(1SH + 1OH) or conditionally NCO: SH = 2:1) for such compositions. Complete curing kinetics of these systems (up to $T_2$ final) is almost the same as of systems with thiokol. The primary (urethane formation) and secondary (allophanate bond formation) reactions with hydroxyl groups look like this:

$$\sim R-N-CH(O) + HO-\overset{\overset{\displaystyle H}{|}}{\underset{\underset{\displaystyle \sim R'\quad SH}{}}{C}}-R'' \longrightarrow \sim R-N-\overset{\overset{\displaystyle (O)}{|}}{C}-O-\overset{\overset{\displaystyle H}{|}}{\underset{\underset{\displaystyle \sim R'\quad SH}{}}{C}}-R''$$

$$2\sim R-N-\overset{\overset{\displaystyle (O)}{|}}{C}-O-\overset{\overset{\displaystyle H}{|}}{\underset{\underset{\displaystyle \sim R'\quad SH}{}}{C}}-R'' + NCO-R''\text{-OCN} \longrightarrow \sim R-N-\overset{\overset{\displaystyle (O)}{|}}{C}-O-\overset{\overset{\displaystyle H}{|}}{C}-R''$$

$$(O)C-\overset{\overset{\displaystyle H}{|}}{N}-R'''-\overset{\overset{\displaystyle H}{|}}{N}-C(O)$$

$$\sim R-N-\overset{\overset{\displaystyle (O)}{|}}{C}-O-\overset{\overset{\displaystyle H}{|}}{C}-R''\sim$$

$T_2$ behavior at the initial curing time indicates compatibility of thiourethane sealant components. Blends of well-compatible oligomers (Thiokol+SKUPFL-100; TPM-2 polymer + SKUPPL-5003) demonstrate the same curing behavior independent on NCO: SH, ratio, while the shape curing kinetics curves of poorly.

Compatible oligomers (Thiokol + SKUPPL-5003; TPM-2 polymer + SKUPFL-100) varies, when the ratio of functional groups is changed. Poorly compatible oligomer pairs demonstrate the maximum rate of curing at the initial period, when prepolymer content is minimal. This period is up to 150 min for thiokol-SKUPPL system and up to 50 min for ТПМ-SKUPFL system. It correlates with results in [72] and seem to due to residual compatibility of components at these ratios (Fig 4.20B; 4.21A).

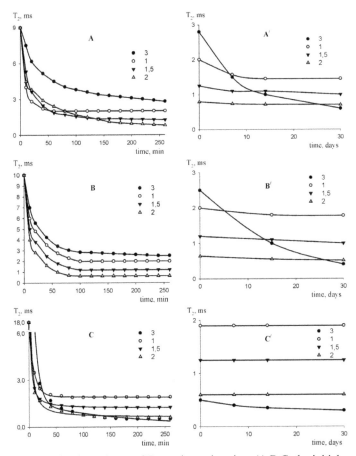

**FIGURE 4.21**  The dependence of $T_2$ on the curing time (A,B,C–the initial period; A′, B′, C′-the final period of curing) time of thiourethane sealants with TPM-2 polymer (A, A′-blend with SKUPFL-100; B, B′-blend with SKUPPL-5003; C, C′-blend with PIC) at the NCO: SH ratio of: ● – 3; ○ – 1; ▼–1.5; △– 2.

Curing rate constant (Fig. 4.22) have been calculated by kinetic dependences (Figs. 4.20, and 4.21). Systems fully compatible with {SKUP-FL-100 + thiokol} (Fig. 4.22A) and {SKUPPL-5003 + TPM-2 polymer} (Fig. 4.22B) have rate constant independent on NCO: SH ratio and the reaction has so-called "pseudofirst" order.

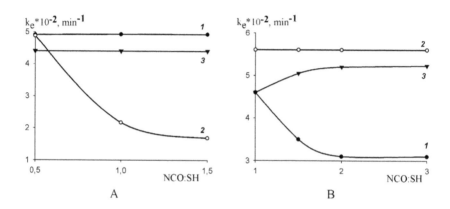

**FIGURE 4.22**   The dependence of curing rate constant for thiourethane sealants on the NCO: SH ratio (A–with thiokol; B–with TPM-2 polymer): 1 – with SKUPFL-100; 2 – with SKUPPL-5003; 3 – with PIC.

It is because this reaction rate constant is a total (resulting) characteristics of several simultaneous reactions: formation of thiourethane, trimerization, formation of thioallophanate bonds as well as urethane formation reaction in the case TPM-2 polymer is used. These reactions occur in the presence of two catalysts: Mannich bases and forming earthen and thiourethane groups, which also have catalytic properties. If low-compatible components are used, a rate constant drops with the increase isocyanate oligomer content. The reaction in the TPM-2 polymer +PIC} system slightly accelerates when NCO: SH ratio increases. It may be thanks to high functionality of PIC as well as low viscosity of TPM-2 polymer that makes interaction of functional groups more probable.

The highest curing rate of {TPM-2 polymer+SKUPPL-5003} system can be explained by increased local concentration of polar oligomeric functional groups (SH and NCO) because of microphase layering of nonpolar blocks of oligomer and polar end fragments [77]. Obtained kinetic dependences are also confirmed by the change kinetics of physic-mechanical properties of thiourethane sealants with time. All the formulations increase their strength with time, however, this increase has different behavior depending on the type of oligomer (Fig. 4.23): the maximum curing rate of systems {Thiokol + prepolymer} is observed when the ratio of reacting groups is equimolar; formulations with the excess of NCO-groups have less strength, than compositions with equimolar NCO: SH ratio at the initial period of curing, that is due to a plasticizing effect of prepolymer excess. Higher final strength is achieved if systems the excess of NCO-groups are cured. It is because of aforementioned participation of NCO-groups in formation of thioallophanate bonds. Formation of transversal bonds results in the increase of strength and decrease of relative elongation. Changing of such parameters as hardness and adhesion to concrete fully correlates with changing of hardness. Formulations with SKUPPL–5003 are different from formulations with SKUPFL–100 by the set of strength properties and by less exceed of the final strength of formulations with the excess of NCO-groups over formulations with equimolar ratio of these functional groups. It is evidently because of less compatibility of SKUPPL–5003 with thiokol comparing SKUPFL–100 and corresponding inhibition of curing, resulting in decreased contribution of secondary bonds to strength.

There is a clear moment of secondary bonding in thiourethane sealants containing thiokol, which falls on approximately 10th day.

Strength (hardness, adhesion to concrete) behavior of {Thiokol+PIC} and TPM-2 polymer-based formulations (Fig. 4.24) is different: the more is the NCO:SH ratio, the better are these characteristics at both initial and final curing stages, maybe because of PIC's low molecular weight and increased functionality of TPM–2 polymer ($f = 4$). The influence of secondary bonds is less in these cases, because structuring is initiated at the very start by primary reactions. Aforesaid is confirmed by the behavior of relative elongation with time, as the excess of isocyanate-containing component acts as a plasticizer in these compositions at the initial stage of curing.

The influence of compatibility on properties of thiourethane sealants ( {SKUPPL–5003 + TPM-2 polymer} system is taken for example) effects in faster getting of strength properties as well as in considerable exceed of strength of formulations with the excess of NCO groups over formulations with equimolar ratio of functional groups comparing formulations on the basis of TPM–2 polymer and SKUPFL–100.

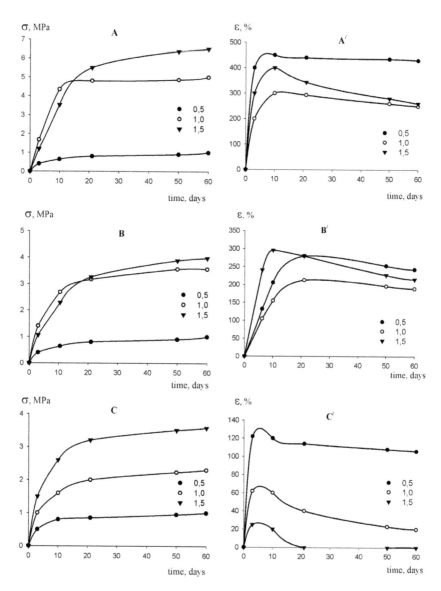

**FIGURE 4.23**   The dependence of strength at rupture (A,B,C); and relative elongation at rupture (A′,B′,C′) of thiourethane sealants based on SKUPFL-100 (A,A′), SKUPPL-5003 (B,B′), PIC (C,C′) and liquid thiokol (thiourethane content is 30 mass parts, OM-3 content is 0,04 mass parts) on curing time (23°C); NCO:SH: ● – 0.5; ○ –1; ▼–1.5

Provided data concerning the rate of how sealants with TPM-2 polymer obtain their properties (Fig 4.24). We can conclude, that secondary reactions start in 3–5 days already that is considerably faster, than for formulations with thiokol (Fig 4.23). It may be due to mutual activation of SH– and OH-groups of TPM-2 polymer [76], as well at catalyzed interaction of NCO groups with SH– and OH-groups and forming thiourethane and urethane groups [72, 75]. However, contribution of secondary reaction is substantially weaker, than for systems with thiokol as in this case the majority of transversal bonds form in primary reactions.

A plasticizing effect of unreacted PSO in all compositions stipulates the highest relative elongation (in comparison with others) of formulation with the lack of isocyanate containing component.

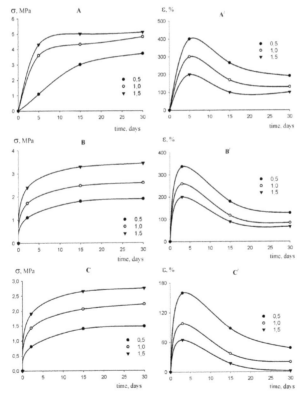

**FIGURE 4.24**   The dependence of strength at rupture (A,B,C); and relative elongation at rupture (A′,B′,C′) of thiourethane sealants based on SKUPFL-100 (A,A′), SKUPPL-5003 (B,B′), PIC (C,C′) and TPM-2 polymer (P-803 – 30 mass parts, OM-3–0.04 mass parts) on curing time (23°C) NCO :( SH+ OH): ● – 0.5; ○–1; ▼ – 1.5.

## 4.5.2   THE DEPENDENCE OF CURING KINETICS AND PROPERTIES OF THIOURETHANE SEALANTS ON THE TYPE AND QUANTITY OF A CATALYST

The influence of catalyst quantity on curing kinetics and properties of thiourethane sealants has been studied for all combinations of PSOs and isocyanate–containing components in question. The behavior of curing rate and properties of cured formulations has been determined to be independent on the type of isocyanate and Therefore, the results for formulations based on SKUPFL–100 are given as an example. The increase of O-M-3 catalyst content in all thiokol-based formulations accelerates curing (viability and $T_2$ values (in 2 h after the start of curing)). Physico-mechanical properties of sealants are almost independent on the dosage of OM–3 (Table 4.17) excluding catalyst dosages leading to either insufficient viability (<10 min.) as a polymer network forms with defects due to high reaction rate; or excessive viability (>6 h), as NCO groups can interact in this case with air moisture releasing carbon dioxide and forming defective structure of cured sealant.

The fact of inessential catalyst influence on final properties of sealants is confirmed by NMR measurements (the estimation of $T_2$ final).

Formulations with TPM–2 polymer can be also cured in the presence of a catalyst. It is explained by mutual activation of mercaptan and hydroxyl groups of TPM-2 polymer [76] and possible initiation of thiourethane formation reaction by forming products, thus TPM–2 polymer obtains and .opportunity of interaction with isocyanate component without a catalyst. IR spectroscopy measurements (Fig. 4.25A) prove, that the reaction between OH- и NCO groups is primary process in systems with TPM-2 polymer and without a catalyst. The figuredemonstrates changing of averaged hydroxyl (–O–H) and amine (=N–H) group signals, as they are in the same area of wave numbers (~3300 $sm^{-1}$). Therefore, IR spectroscopy method is unsuitable for a quantitative estimation of formation (consumption) of groups in TPM–2 polymer-based thiourethane formulations. However, comparison of corresponding areas of IR-spectra of thiokol-based (Fig. 4.25A) (where there is no hydroxyl group and there is clear amplification of amine group's signal) and TPM-2 polymer-based sealants (where the integral (=NH +–OH) signal weakens (Fig. 4.25), reveals the reduction of concentration of hydroxyl groups in the second formulation.

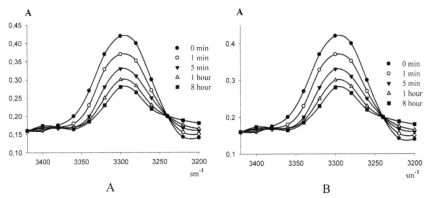

**FIGURE 4.25** The variation of cumulative signals of NH- and OH-groups of thiourethane sealant based on TPM-2 polymer and PIC ((NCO : SH =0.5 : 1) (A – without MB; B–with MB): ●–0 min.; ○–1 min.; ▼–5 min.; Δ – 1 h; ■ – 8 h.

Addition of MBs to the mixture of TPM-2 polymer and isocyanate-containing component accelerates curing process (Table 4.17), as it initiates the thiourethane formation reaction in addition to the urethane formation reaction.

**TABLE 4.17** The Influence of MB Quantity on Curing and Properties of Thiourethane Sealants (Equimolar Ratios, 30 Mass Parts of Carbon Black)

| OM-3, mass parts | $k_e*0,01$, min $^{-1}$ | t, hours | $T_2 2h$ | $T_2$ final | $S_{conv}$, MPa | $e_{rel}$, % |
|---|---|---|---|---|---|---|
| Thiokol + SKUPFL-100 | | | | | | |
| 0.01 | 1.75 | 7 | 0.721 | 0.325 | 3 | 150 |
| 0.025 | 3.3 | 3 | 0.608 | 0.323 | 4.7 | 330 |
| 0.04 | 5.2 | 1 | 0.514 | 0.321 | 4.8 | 320 |
| 0.07 | 6.5 | 0.3 | 0.422 | 0.317 | 5 | 300 |
| 0.1 | 8.1 | 0.1 | 0.387 | 0.311 | 3.5 | 200 |
| TPM-2 + SKUPFL-100 | | | | | | |
| 0 | 2.1 | 6 | 1.56 | 0.76 | 3.5 | 220 |
| 0.04 | 3.5 | 3 | 1.21 | 0.71 | 3.62 | 240 |
| 0.07 | 4.4 | 1.5 | 1.03 | 0.7 | 3.68 | 240 |

Figure 4.26 represents kinetics of changing of NCO-groups concentration in formulations of {TPM-2 polymer + PIC} with and without MBs for

estimation of contributions of urethane formation and thiourethane formation reactions. Concentration drop is more intense for formulations with
MBs indicating higher rate of reaction between mercaptan and isocyanate
groups, than of reaction between hydroxyl and isocyanate groups.

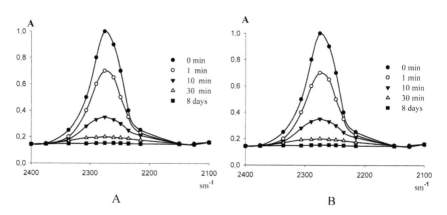

**FIGURE 4.26**    Changing of the NCO-group signal from thiourethane sealant with TPM-2
polymer and PIC (NCO:SH =0,5 : 1) (A – without OM; B – with OM).

Figure 4.25 (A, B) demonstrates, that addition of MB to TPM-2–based
formulations results in less weakening of averaged signal from OH- and
NH-groups, than for formulations without MB, indicating –SH + OCN–
reactions and, therefore, decreased contribution of hydroxyl groups to a
curing reaction. It can be also seen on a lower spectral line (Fig. 4.25A,
B), which corresponds to the complete curing state. The formulation with
MB has this signal a bit stronger, indicating higher residual concentration
of hydroxyl groups. However, reactions of urethane and thiourethane formation occur with a notably high rate that confirms a possibility of mutual
activation of SH and OH-groups.

Final properties of formulations with TPM-2 polymer do not depend
on the content of MB indicating a promoting effect of a catalyst on curing
of thiourethane formulations and their further structuring. It is confirmed
by physic-mechanical tests and $T_2$ values (Table 4.17).

### 4.5.3 THE INFLUENCE OF THE NATURE AND THE RATIO OF COMPONENTS ON PROPERTIES OF THIOURETHANE SEALANTS

Compatibility of components, curing kinetics and final properties of sealants mainly depend on the nature of used oligomers, their functionality and molecular weight.

Table 4.18 demonstrates data on final properties of sealants obtained from reactions of PSO with prepolymers of various nature. Comparison of curing rates (data on the viability (t) and the curing rate constant ($k_e$) obtained from NMR measurements fully correlate with each other) of thiokol and TPM-2 polymer-based sealants leads to a conclusion, that formulations with TPM-2 polymer show much more active transition into the cured state, other things being equal (even in the presence of a filler, which inhibits interaction of TPM-2 polymer with isocyanate-containing component).

TABLE 4.18 The Properties of Thiourethane Sealants (Equimolar Ration of Functional Groups; 30 Mass Parts of Thiourethane; 0,04 Mass Parts of OM-3)*

| PSO+isocyanate system | t, hours | s conv., MPA | $e_{relative}$, % | $A_{concrete}$, MPa | Q water, % | Q toluene, % | $k_e * 0,01$, $min^{-1}$ | $T_2$ final, ms | $n_{eff} \cdot 10^4$ mole/sm$^3$ |
|---|---|---|---|---|---|---|---|---|---|
| Thiokol+PFL (100 + 55) | 2.0 | 5.0 | 250 | 2.1 | 11.4 | 95 | 5.2 | 0.25 | 3.8 |
| Thiokol+PPL (100 + 152) | 6.0 | 3.0 | 170 | 1.8 | 9.8 | 146 | 2.2 | 0.43 | 3.5 |
| Thiokol+PIC (100 + 11) | 2.5 | 2.3 | 20 | 2.5 | 6.2 | 30 | 4.5 | 0.22 | 5.6 |
| TPM +PFL (100 + 110) | 2.2 | 3.7 | 240 | 2.1 | 13.4 | 240 | 5.8 | 0.71 | 3.1 |
| TPM +PPL (100 + 304) | 1.5 | 2.2 | 110 | 1.3 | 11.8 | 60 | 6.2 | 0.51 | 4.6 |
| TPM +PIC (100 + 22) | 1.0 | 1.8 | 30 | 1.6 | 10.9 | 100 | 7.1 | 0.63 | 4.8 |

SH:NCO ratio is 1:2 for compositions with TPM–2 polymer (SH+OH):NCO=1:1, as TPM–2 polymer contains one SH-group per one OH-group.

It is explained by the presence of secondary hydroxyl groups in TPM-2 polymer, which are determined to be able to interact with NCO-groups even without a catalyst.

Achieved degree of curing is slightly higher for formulations with liquid thiokol, than for blends with TPM–2 polymer according to the final value of nuclear spin-spin relaxation time ($T_{2final}$) as well as to the effective density of transversal bonds. Among other isocyanates, PIC demonstrates the highest degree of curing. If TPM-2 polymer is used, formulations based on SKUPPL have the density of network no less, than formulations with PIC. Sealants with TPM-2 polymer or SKUPFL-100 are characterized by a relatively low value of neff, probably because of their insufficient compatibility as well as lower functionality (f=2), than SKUPPL (f≈3) and PIC (f≈2.5).

Formulations with TPM–2 polymer absorb water slightly higher, than systems with thiokol. However, this difference is not as large as for thiokol and TPM-2–containing sealants, cured by manganese dioxide [78]. It is evidently because hydroxyl groups, providing cured formulations with increased hydrophobicity, stay unshielded during curing of TPM-2 polymer by oxidative curing agents.

Maximum strength, elastic and adhesive properties are characteristic for sealants with SKUPFL–100. It is probably due to a contribution of a prepolymer to the properties of thiourethane sealant. It should be noted, that SKUPFL–100 has low molecular weight of 1350. Therefore, the content of thiourethane groups in a sealant with SKUPFL–100 is several times higher than in sealants with SKUPPL–5003 (the equivalent molecular weight of SKUPFL-100 is ~ 675, while its value for SKUPPL-5003 is ~1700).

Thermal resistance of studied formulations is first of all influenced by the density of polymer network and the nature of oligomeric main chain (Table 4.19).

**TABLE 4.19** Thermal Properties of Thiourethane Sealants (The Ratio of Functional Groups is Equimolar; 30 Mass Parts of Thiourethane; 0.01% of OM-3)

|              | Liquid Thiokol | TPM-2 polymer |
|--------------|----------------|---------------|
| SKUPFL-100   | 180/–60        | 165/–60       |
| SKUPPL-5003  | 190/–60        | 200/–60       |
| Polyisocyanate | 210/–60      | 220/–60       |

* the numerator is the starting temperature of thermal destruction, °C; the denominator is the glass transition temperature, °C.

The Table 4.19 demonstrates that the decrease of compatibility of components reduced the starting temperature of thermal destruction: changing of liquid Thiokol to TPM–2 polymer in systems with SKUPFL–100 leads to the decrease of the starting temperature of thermal destruction; and vice versa for systems based on SKUPPL–5003. Formulations with PIC are characterized by the best thermo mechanical characteristics independent on PSO type because of increased concentration of solid blocks in such compositions.

There is the research concerning the influence of PSO and oligoisocyanate ratio on the properties of thiourethane sealants [79]. The ratio of oligomers in systems with liquid thiokol was changed in the following limits, considering the content of end groups (molar): NCO:SH = 0.5÷1.0÷1.5 : 1, while these limits for formulations with TPM–2 polymer were: NCO:SH = 1÷2÷3 : 1, considering constituent OH-groups. The variation of strength of sealants depending on the ratio of components is represented in Fig 4.27. It is obviously, that sealants have reduced strength with the lack of isocyanate groups independently on PSO and isocyanate component types. However, their relative elongation is increased due to a plasticizing effect of PSO excess. It also confirmed by values of neff., $T_2$ final and swelling in various media (Tables 4.20, and 4.21).

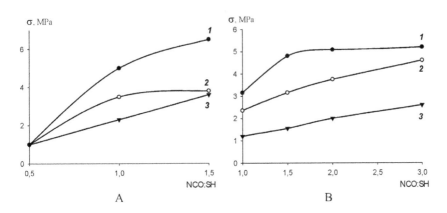

**FIGURE 4.27**    The dependence of strength at rupture of thiokol (A) and TPM–2 polymer (B)–based thiourethane sealants on the NCO : SH ratio (thiourethane content is 30 mass parts; OM–3 content is 0,04 mass parts): 1 – SKUPFL–100; 2 – SKUPPL–5003; 3 – polyisocyanate.

**TABLE 4.20** The dependence of properties of thiokol–based thiourethane sealants (thiourethane content is mass parts; OM–3 content is 0,01 mass parts) on the NCO:SH ratio

| Isocyanate | NCO:SH | $T_2$ final, ms | $n_{eff.}*10^4$, mole/sm$^3$ | $n_{chem.}*10^4$, mole/sm$^3$ | $Q_{water}$, % weight | $Q_{toluene}$, % weight |
|---|---|---|---|---|---|---|
| | 0.5 | 0.546 | 2.0 | 0.8 | 15 | 150 |
| PFL | 1 | 0.25 | 3.8 | 1.4 | 11.4 | 95 |
| | 1.5 | 0.146 | 4.3 | 1.6 | 12.2 | 115 |
| | 0.5 | 0.834 | 2.1 | 0.5 | 15.2 | 200 |
| PPL | 1 | 0.421 | 3.5 | 0.8 | 9.8 | 145 |
| | 1.5 | 0.271 | 4.1 | 1 | 10.1 | 160 |
| | 0.5 | 0.51 | 2.7 | 1.1 | 7.8 | 57 |
| PIC | 1 | 0.226 | 5.6 | 2 | 6 | 30 |
| | 1.5 | 0.133 | 6.5 | 2.2 | 6.5 | 34 |

Strength of all studied systems grows, when the content of NCO groups increases. The behavior of this process depends on the nature of oligomers. It can be seen, that if oligomers are well–compatible, strength increases monotonically with isocyanate content. If oligomers are poorly compatible, the growth of strength is much slower (thiokol + SKUPPL–5003; TPM–2 polymer + SKUPFL–100).

Minimal values of water absorption and swelling in toluene are typical for formulations with NCO : SH = 1 : 1 (for systems containing thiokol) and NCO : SH = 1.5 : 1 (for systems containing TPM–2 polymer). The presence of unreacted thiokol in a cured sealant (with the shortage of NCO) makes sealant less resistant to water and solvents. If there is excess of isocyanate, unreacted NCO groups can react with air moisture releasing carbon dioxide in the following reaction [70]:

$$\sim RNCO + H_2O + OCNR'\sim \longrightarrow \sim R\overset{H}{-}N-\overset{(O)}{C}-\overset{H}{N}-R'\sim + CO_2\uparrow$$

the related coating will be defective due to included gas and will swell more in water or toluene, than formulations with equimolar NCO : SH ratio.

**TABLE 4.21** The dependence of properties of TPM–2 polymer–based thiourethane sealants (thiourethane content is mass parts; OM–3 content is 0,01 mass parts) on the NCO:SH ratio

| Isocyanate | NCO:SH | $T_2$ final, ms | $n_{eff.}*10^4$, mole/ sm$^3$ | $n_{chem}.* 10^4$, mole/sm$^3$ | $Q_{water}$, % weight | $Q_{tolu-ene}$, % weight |
|---|---|---|---|---|---|---|
| PFL | 1 | 1.45 | 1.2 | Break | 15.8 | Break |
| | 1.5 | 1 | 2.0 | 0.5 | 12.9 | 300 |
| | 2 | 0.71 | 3.1 | 0.8 | 13.4 | 240 |
| | 3 | 0.64 | 3.5 | 0.9 | 14.2 | 280 |
| PPL | 1 | 1.78 | 3.0 | 1.1 | 13.7 | 200 |
| | 1.5 | 1 | 3.7 | 1.8 | 9.8 | 100 |
| | 2 | 0.51 | 4.6 | 2.2 | 11.8 | 60 |
| | 3 | 0.4 | 4.9 | 2.3 | 12.4 | 120 |
| PIC | 1 | 1.9 | 3.1 | 0.7 | 12.8 | 179 |
| | 1.5 | 1.21 | 4.1 | 1.4 | 8.4 | 140 |
| | 2 | 0.63 | 4.8 | 1.8 | 10.9 | 100 |
| | 3 | 0.34 | 5.0 | 2.0 | 11.6 | 130 |

The growth of NCO: SH ratio naturally increases neff. and nchem. values, andnchem. grows faster for compositions with high compatibility.

PSO-isocyanate component ratio stipulates adhesive properties of a sealant, in particular, to concrete [79] (Fig 4.28). Adhesion of completely cured sealants made of SKUPPL–5003 or polyisocyanate is maximal at equimolar ratio of functional groups; further growth of content of NCO groups has less effect on the increase of adhesion excluding sealants based on thiokol and SKUPFL–100 which become more adhesive, if the content of NCO groups is above equimolar. The tear from concrete property of all the formulations with the shortage of NCO groups is of cohesive type and is limited by sealant's strength. When the content of NCP groups is close to equimolar to SH groups, as well as to hydroxyl groups for TPM–2 polymer, the tear type changes to adhesive as the strength of formulations exceeds adhesion to concrete. The analysis of breaking pattern indicates, that "true" adhesion values are not determined, because if the NCO : SH ratio is above equimolar, sealant tears from the concrete surface with the surface (weak) concrete layer.

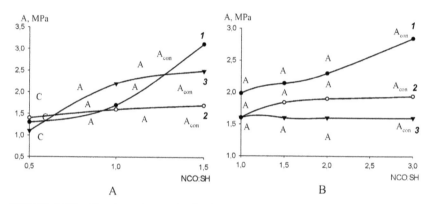

**FIGURE 4.28** The dependence of adhesion to concrete of thiourethane sealants with thiokol (A) and TPM–2 polymer (B) on the NCO:SH ratio (thiourethane content is 30 mass parts; OM–3 content is 0.04 mass parts): 1 – SKUPFL–100; 2 – SKUPPL–5003; 3– polyisocyanate. C – cohesive fracture; A – adhesive fracture; $A_{con}$ – adhesive rupture with concrete crump.

## KEYWORDS

- **epoxy resin**
- **liquid thiokol vulcanizates**
- **thiourethane sealants**
- **unsaturated polyesters**

## REFERENCES

1. Averko-Antonovich, L. A., Kirpichnikov, P. A., & Smyslova, R. A. (1983). Polysulfide Oligomers & Related Sealants (in Russian). Leningrad: Himija, 128 p.

2. Smyslova, R. A. (1974). Liquid Thiokol Sealants (in Russian) Moscow: CNIIT Jeneftehim, 83 p.
3. Smyslova, R. A., Kotljarova, S. V. (1976). Handbook on Rubber Sealing Materials (in Russian) Moscow: Himija 72 p.
4. Mudrov, O. A., Savchenko, I. M., & Shitov, V. S. (1982). Handbook on Elastomer Coatings & Sealants in Shipbuilding Leningrad: Sudostroenie 184 p.
5. Labutin, L. P. (1982). Anticorrosive and Sealing Materials Based on Synthetic Rubbers (in Russian). Leningrad: Himija 213 p.
6. Nefed'ev, E. S., Hakimullin, Ju. N., Polikarpov, A. P., & Averko-Antonovich, L. A. (1986). IzvestijaVUZov (in Russian), 29(1), 97–100.
7. Muhutdinova, T. Z., Shahmaeva, A. K., Gabdrahmanov, F. G., & Sattarova, V. M. (1980). Kauchukirezina (in Russian), 1, 12–15.
8. Barron, B. (1955). Rub & Plaste Age, 36(1), 26–28.
9. Polikarpov, A. P. (1982) Technical Sciences Candidate's Dissertation (in Russian), Kazan, KCTI.
10. Polikarpov, A. P., Averko-Antonovich, Ju. O., Romanova, G. V., & Krasil'nikov, A. P. (1982). Kauchuk, irezina (in Russian), (2) 27–28.
11. Suhanov, P. P. (2003). Chemical Sciences Doctor's Dissertation (in Russian), Kazan, KSTU.
12. Paken, A. M. (1962). Epoxy Compounds & Epoxy Resins (in Russian), Leningrad, Goshimizdat, 963 p.
13. Averko-Antonovich, L. A., Kirpichnikov, P. A., Materials of International Conference on Caoutchouc& Rubber (in Russian). Kiev P.44–51.
14. Minkin, V. S. (1975). Chemical Sciences Candidate's Dissertation (in Russian) Kazan' KCTI.
15. Minkin, V. S., Averko-Antonovich, L. A., & Kirpichnikov, P. A. (1975). Vysokomolek-Soed. (in Russian). Series B, 17(1), 26–29.
16. Kuska, H., Rodzhers, M. (1970). EPR of Transition Metal Complexes (in Russian), Moscow, 82 p.
17. May, C., Tanaka, Y. (1973). Epoxy Resins, N. Y., 801p.
18. Popl, D., Shnejder, V., & Bernstejn, G. (1962). High Resolution NMR Spectra (in Russian), Moscow, 208 p.
19. Elchueva, A. D. (1989). Chemical Sciences Candidate's Dissertation (in Russian), Kazan, KCTI.
20. Polikarpov, A. P., Averko-Antonovich, L. A., Romanova, G. V.,& Krasil'nikova, A. P. (1982). Kauchukirezina (in Russian). (2) 27–29.
21. Petrov, O. V., Nefed'ev, E. S., & Kadirov, M. K. (1998). Kauchukirezina (in Russian). (5) 13–19
22. Ashihmina, L. I., Jamalieva, L. N., Nefed'ev, E. S., & Averko-Antonovich, L. A. (1988). Kauchukirezina (in Russian). (12) 19–21.
23. Polikarpov, A. P., Averko-Antonovich, L. A., et al, & IzvestijaVUZov, (1982). "Chemistry & Chemical Technology" series, 25(11) 1388–1392.
24. Nefed'ev, E. S. (1991). Chemical Sciences Doctor's Dissertation (in Russian), Kazan, 233 p.
25. Nefed'ev, E. S., Ashihmina, L. I., Ismaev, I. Je., Kadirov, M. K., Averko-Antonovich, L. A., & Il'jasov, A. V. (1989). Dokl. AN SSSR (in Russian), 304(5) 1181–1184.

26. Kirpichnikov, P. A., Il'jasov, A. V., & Kadirov, M. K. et al. (1986). Izv AN SSSR. The Chemical Series (in Russian), (12) 2824.
27. Kirpichnikov, P. A., Il'jasova, A. V., & Kadirov, M. K. et al. (1987). Dokl. AN SSSR (in Russian). 294(4) 910–913.
28. Suhanov, P. P., Khakimullin, Yu. N., Kostochko, A. V., & Dzhanbekova, L. R. (1996). Zhurnalprikladnojhimii (in Russian) (1) S.124–126.
29. Khakimullin, Yu. N., Gafurov, F. Sh., & Egorova, N. K. (1986) i dr. Abstracts of a National seminar "Prospective Elastomer & Elastomer-oligomer compositions for rubber products" (in Russian). Moscow: CNIITJenefte him. 32 – 33.
30. Suhanov, P. P., Minkin, V. S., Averko-Antonovich, L. A., Gafurov, F. Sh., & Khakimullin, Yu. N. (1989). Acryl oligomers: Synthesis, Properties & Application." Book of Articles from Institutes Gor'kij S. 42–46.
31. Khakimullin, Yu. N., Gafurov, F. Sh., Egorova, N. K., & Nefed'ev, E. S. (1985). Abstracts of the National Conference "Quality Improvement & Resource-saving Technologies for the Rubber Industry" (in Russian). Jaroslavl 17.
32. Khakimullin, Yu. N., Idijatova, A. A., Kurkin, A. I., Valeev, R. R., & Gubajdullin, L. Ju.,( 1998). Abstracts of 16th Mendeleev Congress on Pure & Applied Chemistry" (in Russian) Moscow, (2) 183.
33. Slonim, I. Ja., Urman, Ja. G. (1982). NMR Spectroscopy of Heterochain Polymers (in Russian), Himija, M., 300 p.
34. Mirakova, T. Ju., Nefed'ev, E. S., Ismaev, I. Je., Averko-Antonovich, L. A., & Il'jasov, A. V. (1990). Plastmassy (in Russian). (3). P. 27–28.
35. Mirakova, T. Yu., Nephedijev, E. S., Ismaev, I. E., Averko-Antonovich, L. A., & Il'yasov, A. V. (1989). Magn. Resonance Polym.: 9th Spec. Colleg. Ampere. Prague: [Pap. on] 12th Discuss. Conf. Prague Meet. Macromol.: Program. Prague. (1989) 105–106.
36. Shljahter, R. A., Novosjolok, F. B., & Apuhtina, N. P. (1971). Kauchukirezina (in Russian) (2), 36–37.
37. Apuhtina, N. P., Shljahter, R. A., & Novosjolok, F. B. (1957). Kauchukirezina (in Russian) (6), 7–11.
38. Bertozzi, E. R. (1968). Rubb. Chem. and Technol. 41(1), 114–160
39. Brown, H. P. (1957). Carbohylised elastomers. Rubb. Chem. and Technol. 30(5), R.1347–1386
40. Khakimullin, Yu. N., Kurkin, A., Ioffe, D., Galimzyanov, R., & Liakumovich, A. (1999). International Conferences on Polymer Characterization. University of North Texas, Denton, Second Announcement of the 7th POLYCHAR, January 5–8, Book of abstracts 41
41. Bloh, G. A. (1972). Organic Promoters for Rubber vulcanization (in Russian). Leningrad, Himija, 560 p.
42. Khakimullin, Yu. N., Kurkin, A. I., Gafurov, F. Sh., & Liakumovich, A. G. (2000). Zhurnalprikladnojhimii (in Russian), 73(3), 501–504.
43. Tager, A. A. (1978) Physical Chemistry of Polymers, 3rd Edition (in Russian). Moscow, "Himija," 339 p.
44. Kurkin, A. I. (2001) Technical Sciences Candidate's Dissertation (in Russian), Kazan, KSTU.

45. Muhutdinova, T. Z., Averko-Antonovich, L. A., Prokudina, K. N., & Kirpichnikov, P. A. (1973). KSTU Transactions (in Russian). Issue 50 P. 141–145.
46. Khakimullin, Yu. N., Kurkin, A. I., Liakumovich, A. G., & Ionov, Ju. A. (2001). Kauchukirezina (in Russian), 4 P.22–25.
47. Muhutdinova, T. Z., Averko-Antonovich, L. A. (1971). Kauchukirezina (in Russian) 12, 10–13.
48. Li, T. P. S. (1995). Kauchukirezina (in Russian) 2, 9–13.
49. Averko-Antonovich, L. A., Kirpichnikov, P. A., & Romanova, G. V. (1969). KSTU Transactions (in Russian), Issue 40, Part 2, 54–62
50. Averko-Antonovich, L. A., Kirpichnikov, P. A., Tazetdinova, N. N., et al. (1973). KSTU Transactions (in Russian). Issue 40. P. 47–53.
51. Patent USA3813368, (1974) MKI S 08 G 75/04 Polysulfide compositions (in Russian).
52. Romanova, G. V., Rubanov, V. E., & Averko-Antonovich, L. A. (1978). Abstracts of VDNH USSR (in Russian), Moscow, CNIIT Jeneftehim, 15.
53. Berlin, A. A., Kefeli, T. Ja., & Korolev, G. V. (1967). Polyesteracrylates (in Russian). Moscow: Nauka. −270 p.
54. Minkin, V. S., Romanova, G. V., & Averko-Antonovich, L. A., et al. (1982) Vysokomolek. Soed. (in Russian), Vol. 24B, 7, 806–809.
55. Muhutdinov, M. A., Khakimullin, Yu. N., Gubajdullin, L. Ju., & Liakumovich, A. G. (1996). Abstracts of 3rd Russian Theoretical & Practical Conference on Rubber (in Russian). Moscow. P.244–245.
56. Vahonin, A. P., Gubajdullin, L. Ju., Muhutdinov, M. A., Skryleva, V. V., & Khakimullin, Yu. N. (1996). Abstracts of Theoretical & Practical Conference. "Present State and Prospectives of synthetic Rubbers, Polysulfide Oligomers & Their Dericatives" (in Russian). –Kazan. P.18
57. Muhutdinov, M. A., Khakimullin, Yu. N., Gubajdullin, L. Ju., & Liakumovich, A. G. (1997). Abstracts of 4th Russian Theoretical & Practical Conference on Rubber (in Russian), Moscow. 281–282.
58. Muhutdinov, M. A., Khakimullin, Yu. N. (1997). Collected papers of theoretical & practical conference "Production & Consumption of Sealants and Other Construction Materials: Present State and Prospectives (in Russian). Kazan. 65.
59. Muhutdinov, M. A., Khakimullin, Yu. N., Gubajdullin, L. Ju., & Liakumovich, A. G. (1998). Kauchukirezina (in Russian), (3) 33–35.
60. Khakimullin, Yu. N., Valeev, R. R., Minkin, V. S., & Liakumovich, A. G. (1999). Abstracts of 5th International Conference on Intensification of Petrochemical Processes "Neftehimija 99" (in Russian). Nizhnekamsk, 114–115.
61. Valeev, R. R., Owepkov, O. Je., Khakimullin, Yu. N., Minkin, V. S., Gubajdullin, L. Ju., & Liakumovich, A. G. (2000). Abstracts of 2nd all Russian Kargin symposium "Polymer Chemistry & Physics in Early twenty-first century" (in Russian) Chernogolovka 1–65.
62. Khakimullin, Yu. N., Valeev, R. R., Gubajdullin, L. Ju., Minkin, V. S., Owepkov, O. V., Liakumovich, A. G., Deberdeev, R. Ja., & Zaikov, G. I. (2002). Curing of Thiokol Sealants in the Presence of Unsaturated Compounds (in Russian) Plastmassy. (7) 33–36.
63. Patent 2153517 Russia, MKI S 09 K3/10.

64. Askadskij, A. A., Matveev, Ju. I. (1983). Chemical Structure & Physical Properties of Polymers Moscow Himija 248 p.
65. Van-Krevelen, D. V. (1976). Properties & Chemical Structure of Polymers. Moscow Himija, 416 p.
66. Rahmatullina, G. M. (1982). Chemical Sciences Candidate's Dissertation (in Russian), Kazan, KCTI.
67. Sharifullin, A. L. (1991). Chemical Sciences Candidate's Dissertation (in Russian), Kazan, KCTI.
68. Khakimullin, Yu. N. (2003). Technical Sciences Doctor's Dissertation (in Russian), Kazan, KSTU
69. Apuhtina, N. P., Novoselok, F. B., Kurovskaja, L. S., & Ternavskaja, G. K. (1970). Synthesis & Physical Chemistry of Polymers (in Russian), Kiev, NaukovaDumka, (6), 141–143.
70. Saunders Dzh, H., Frish, K. K. (1968). Chemistry of Polymers (in Russian) Moscow Himija 470 p.
71. Kurkin, A. I., Khakimullin, Yu. N., Petrov, O. V., Nefed'ev, E. S., & Liakumovich, A. G. (2000). Book of articles of 7th all Russian Conference. "Structure & Dynamics of molecular systems" (in Russian) Moscow, 504–506.
72. Minkin, V. S., Averko-Antonovich, L. A., Kirpichnikov, P. A., & Suhanov, P. P. (1989). Vysokomolek. soed. (in Russian) 31(2), 238–251.
73. Suhanov, P. P., Averko-Antonovich, L. A., Elchueva, A. D., & Dzhanbekova, L. R. (1996). Kauchukirezina (in Russian) (3), 28–32.
74. Habarova, E. V., Saharova, M. A., Harakterova, L. V., & Morozov, Ju. L. (1997). Kauchukirezina (in Russian) (3), 17–21.
75. Elchueva, A. D., Tabachkov, A. A. (2002). Prikladnajahimija (in Russian) 75(8) 1338–1340.
76. Idijatova, A. A. (1999). Technical Sciences Candidate's Dissertation (in Russian), Kazan, KSTU
77. Kurkin, A. I., Khakimullin, Yu. N., & Liakumovich, A. G. (2000). Kauchukirezina (in Russian) (5), 33–36.

**CHAPTER 5**

# THE INFLUENCE OF FILLERS ON THE PROPERTIES OF POLYSULFIDE OLIGOMER SEALANTS

## CONTENTS

## 5.1   THE INFLUENCE OF THIXOTROPIC ADDITIVES ON THE PROPERTIES OF SEALANTS

Thixotropicity, that is sealant's resistance to flowing down a vertical surface, is one of the main technological properties, required from sealants designed for sealing of interpanel joints in house building, which stipulates their application capabilities. [1]. The viscosity and thixotropicity of compositions based on high molecular weight polymers or oligomers of various nature is known to be increased by addition of silica, montmorillonites (bentonites) as well as by such hydroxyl-containing compounds as ethylene glycol (EG) and glycerol [2–4]. However, there is no systematic research in the selection of additives and their influence on viscous and thixotropic properties. Therefore, a study has been carried out concerning the influence of silica (aerosils A-175 and A-300, carbon white BS-120 and caoline) and various glycols (water, EG, diethyleneglycol (DEG), glycerol) on the viscous and thixotropic properties of filled (natural chalk −100 mass parts) and unfilled sealants based on thiokol-containing oligomer (TPM-2 polymer) and liquid thiokol. These sealants were cured by manganese dioxide. Oxidant's excess coefficient (in moles) was 3 in relation to end SH-groups.

As Fig. 5.1 shows, addition of A-175 aerosil to TPM-2 polymer and liquid thiokol results in substantial increase of viscosity. This effect is much stronger in TPM-2 polymer systems without chalk, than in formulation with liquid thiokol. The explanation can be the difference of PSOs viscosities. However, this effect is first of all stipulated be the presence of hydroxyl groups in TPM-2 polymer chain, which are able to form hydrogen bonds with aerosil's hydroxyl groups [5–8].

TPM-2 polymer is based on a simple PE, which is laprol–4503, produced by polymerization of propylene oxide in the presence of basic catalyst. Each of its monomer links contains oxygen in an ether group:

$$\left(\begin{matrix} H2 & H2 \\ C\!-\!C\!-\!O \\ | \\ CH3 \end{matrix}\right)_{n}$$

Each monomer link of thiokol contains ether oxygen too as well as sulfur: $(\sim CH2\!-\!CH2\!-\!O\!-\!CH2\!-\!O\!-\!CH2\!-\!CH2\!-\!SS\sim)_{n}$. Oxygen atoms of ether groups can form hydrogen bonds with aerosols hydroxyl groups.

However, experimental results demonstrate (Fig 5.1B, curve 2), that these atoms are not inclined to form hydrogen bonds and their contribution can be ignored, when obtained results are analyzed. This explanation is also suitable for laprol's ether group.

Indeed, the calculated quantity of hydroxyl groups in formulation based on TPM-2 polymer is 0.967 grams per 100 grams of oligomer that is equal to addition of extra 1.87 grams of ethylene glycol. It seems to explain a substantial increase of viscosity by caused by aerosil only (Fig. 5.1A, curve 2).

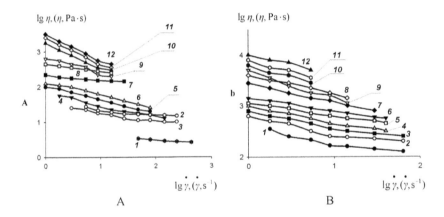

**FIGURE 5.1**   Viscosity-shear rate dependence for formulations with chalk (– – –) and without chalk (—): A – with TPM-2 polymer (A–175 = 2 mass parts, glycol = 5 mass parts); B – on the basis of liquid thiokol (A–175 = 2 mass parts): 1 – reference sample; 2.8 – A-175; 3,9 – glycerol; 4.10 – DEG; 5,11 – EG; 6,12 – water; 7 – a formulation with chalk.

From the point of structuring, aerosils become more effective with glycols, such as water, EG, DEG and glycerol, which can form hydrogen bonds with aerosil and PSO thanks to high concentration of hydroxyl groups.

The analysis of results (Fig 5.1) has shown, that glycols in unfilled formulations with TPM-2 polymer or liquid thiokol (curves 1–6) demonstrate the following efficiency to viscosity increase:

$$H_2O > EG > DEG \geq glycerol$$

Considering the fact, that glycerol has one secondary hydroxyl group (which is substantially less active); there is a correlation between the content of primary hydroxyl groups in above glycols and their efficiency to viscosity increase. If 5 mass parts of DEG or glycerol are added to formulations based on TPM-2 polymer per 100 mass parts of oligomer, the viscosity increase effect in observed only at low shear rates. They act as plasticizers at high shear rates and reduce system viscosity below its initial value.

Such dependences can be observed in system filled with chalk and added glycols (Fig. 5.1A, curves 7–12). Estimations of glycol effect on filled formulations based on TPM-2 polymer (Fig. 5.1A) show, that the viscosity increase effect is stronger in these systems, than in unfilled ones.

Figure 5.1A demonstrates, that the viscosity of systems with added glycol increases faster, than viscosity of the same formulations without glycols. It can be explained by the increase of thixotropicity effect because of aforementioned presence of TPM-2 polymer fragments able to interact with aerosils and polar glycols.

The increase of viscosity caused by addition of glycols is less active in filled compositions based on liquid thiokol (Fig. 5.1B, curves 7–12).

It is interesting to study the variation of viscosity with the increase of glycol content. Addition of up to 10 mass parts of water or EG to unfilled formulations based on TPM-2 polymer causes viscosity increase. The same effect is observed for 2 and 5 mass parts of added glycerol and DEG correspondingly.

There is more intense increase of viscosity in systems filled with chalk (Fig. 5.2), than in unfilled ones. The highest viscosity value is achieved in formulations containing 5–7 mass parts of glycol. Further increase of content of glycols does not lead to a substantial change of viscosity. The same effects are observed in filled formulation based on liquid thiokol, however, they are much weaker.

Application opportunities of caoline have been also studied in addition to the study of silica and glycol effects on viscosity and thixotropicity of PSO-based sealants (Fig. 5.3). Kaolin is a hydrated aluminum silicate, which consists of chemically bound silicon dioxide and hydrated aluminum dioxide. It can provide sealants with Thixotropic properties too, depending on their composition.

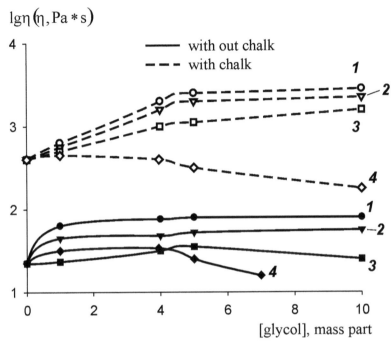

**FIGURE 5.2**   Viscosity variations of formulations with chalk (– – –) and without chalk (—) based on TPM-2 polymer with various content of glycols (the addition of A-175 is 2 mass parts) at the shear rate (lgg) = 0.25: 1 – water; 2 – EG; 3 – DEG; 4 – glycerol.

The analysis of results has shown (Fig 5.3), that 100 mass parts of caoline are required t provide systems based on TPM-2 polymer with the viscosity of 100 mass parts of chalk + 2 mass parts of aerosil + glycol combination. Addition of so much of caoline to sealants reduces curing rate (pH of kaolin = 3.0–4.0), which is not always necessary, as well as to deterioration o elastic properties, that is undesirable. Addition of polar glycols increases the viscosity of systems with kaolin. However, this effect is higher, when chalk is used because of bound water in kaolin. Therefore, kaolin can be added to construction sealants only as secondary (additional) filler.

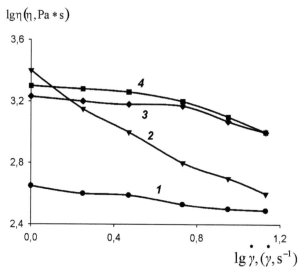

$\lg \eta\,(\eta, \text{Pa} * \text{s})$

$\lg \gamma, (\gamma, \text{s}^{-1})$

**FIGURE 5.3**   The dependence of dynamic viscosity on a shear rate of compositions with TPM-2 polymer (100 mass parts of filler: 1,2 – chalk; 3,4 – kaolin): 1 –A–175 (2 mass parts); 2 – A-175 (2 mass parts) + $H_2O$ (5 mass parts); 3 – no additives; 4 – $H_2O$ (5 mass parts).

Experimental results can be explained by formation of labile network coagulation structures in PSO-silica-glycol system, which are destroyed, when the shear rate is increased and reappear, when it's decreased. This effect is more intense for formulations with TPM-2 polymer for the afore-mentioned reasons (Fig. 5.4).

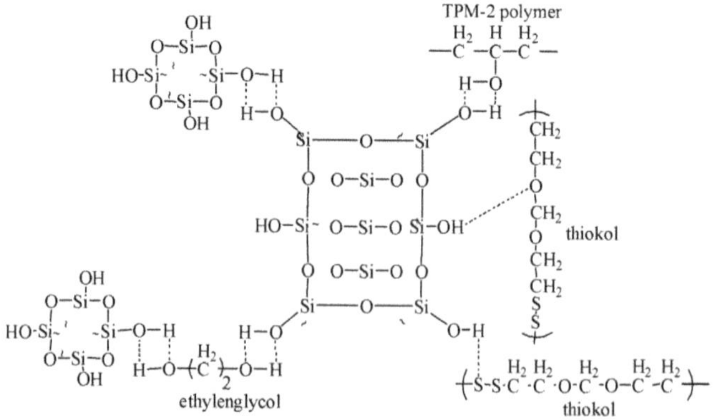

**FIGURE 5.4**   Formation of hydrogen bonds in aerosil-glykol-PSO system.

The analysis of "resistance to fluidity" parameter (Fig. 5.5), which is a common production characteristic of construction sealers (GOST 25621–83) and indicates thixotropicity of sealants, has proved the validity of above conclusions on viscous and thixotropic properties of PSO-based compositions filled with natural chalk (100 mass parts) [1, 6].

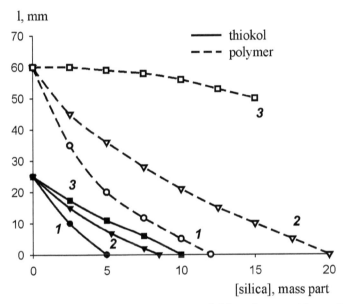

**FIGURE 5.5**    The dependence of resistance to fluidity of compositions with TPM-2 polymer (——) and liquid thiokol (–) on the content of silica: 1 – A-300; 2 – A-175; 3 – BSБC–120.

Figure 5.5 demonstrates that the effectiveness of silica in sealants with TPM-2 polymer depends on their dispersity:

$$A–300 > A-175 > БC–120.$$

It is stipulated by the increase of specific area of oligomer-silica contact and the concentration of polar OH-groups forming hydrogen bonds. Although application of A-300 aerosil is more effective, than of A-175, A-175 aerosil is usually added to blends in practice because fine dispersed aerosils are difficult to be introduced into oligomer systems using existing equipment. BS-120 carbon white is less effective, than it was expected due to less surface concentration of –SiOH-groups and related difference in surface properties. Activity of silica jumps up with addition of water (Fig. 5.6). If 12, 5 mass parts of dry A-300 are required to achieve required effect (Fig. 5.5), than it is 2,5 mass parts with water (Fig. 5.6). Addition of water considerably increases the effectiveness of BS-120 too.

Formulations based on liquid thiokol show the same effects. However, there is less difference between the efficiency of studied additives (Fig. 5.5). БC–120 additive is more active in these systems, than in formulations based on TPM-2 polymers, and in close to the efficiency of aerosils.

The influence of EG, glycerol and EG content on properties of sealants has also been estimated. As Fig. 5.7 shows, formulations with "zero" fluidity containing TPM-2 polymer can be prepared, if the content of glycols in a system is 7–9 mass parts, and they have similar efficiency. Addition of more glycols (up to 70 mass parts of water, up to 50 mass parts EG or DEG and up to 20 mass parts of glycerol) does not influence their thixotropicity and "zero" fluidity property (Fig. 5.7). Further additives of glycols above these dosages deprive formulations of thixotropicity, and they start to flow. We consider these results to be rather interesting as they lead to a conclusion, that additives of glycols below "critical" values do not make a plasticizing effect on sealant and form three-dimensional coagulation structures with aerosil

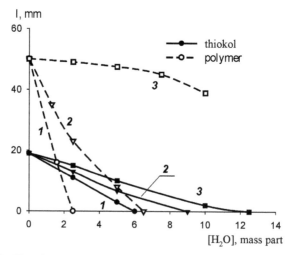

**FIGURE 5.6** The dependence of resistance to fluidity of compositions with TPM-2 polymer (– – –) and liquid thiokol (—) on the content of water (2 mass parts of silica is added): 1 – A-300; 2 – A-175; 3 – БC–120

However, it is also of practical interest: sealants may become cheaper and keep their viscous, thixotropic and performance characteristics, if relatively large quantities (from 30 to 100 mass parts) of EG or DEG is added with simultaneous introduction of chalk.

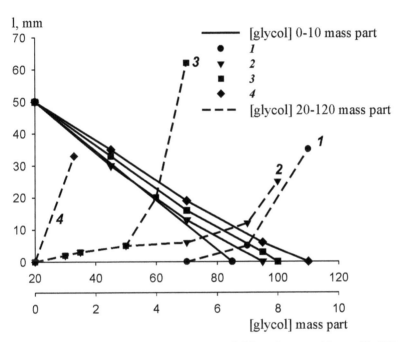

**FIGURE 5.7** The dependence of resistance to fluidity of compositions with TPM-2 polymer on the content of glycol (A–175 = 2 mass parts): 1 – water; 2 – EG; 3 – DEG; 4 – glycerol.

Water is the most effective additive for liquid thiokol formulations (Fig. 5.8). Other additives act as expected: although they reduce viscosity, but the required flow resistance (L=2 mm) is not achieved for the given content of aerosil (2 mass parts).

DEG and glycerol additives start to increase fluidity if their content is at least 20 mass parts, this value is 30 mass parts for EG and 50 mass parts of water (Fig. 5.8).

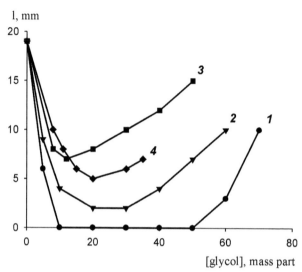

**FIGURE 5.8** The dependence of flow resistance of liquid thiokol formulations on glycol content (A–175 = 2 mass parts): 1 – water; 2 – EG; 3 – DEG; 4 – glycerol.

Rapid increase of this parameter can be explained by two reasons:

1. A plasticizing effect emerges in formulations with large content of glycol.
2. Separation of glycols into an individual phase due to their incompatibility with PSO and resulting increase of fluidity because of "greasing effect." This effect is typical, when glycerol or DEG is used.

Liquid thiokol formulations require more aerosil, than compositions with TPM-2 polymer to achieve required flow resistance. Therefore, the efficiency of viscosity and thixotropicity control by addition of glycols and silica has been determined to depend on the nature of PSO. Formulations with TPM-2 polymer are more inclined to it, than compositions with liquid thiokol. The analysis of experimental data makes it possible to recommend, taking the presence of fillers into account, the required additives of silica and glycols to produce sealants with optimal viscosity and thixotropicity from the viewpoint of their application.

The analysis of experimental results has shown the effect of synergism in silica-glycol system. It is the most intense, if compounds with high concentration of hydroxyl groups are used for glycols. The optimal combination of silica and glycol substantially reduces the content of aerosol required for sustainable viscous properties of sealants that is cost-beneficial due to high prices for aerosols. However, a final selection of silica and glycol should be made considering their influence on the curing rate and physic-mechanical properties of sealants.

As 5–10 mass parts of silica fillers with synergists (glycols) are required for the control of viscosity and required thixotropic properties, we can expect a substantial effect of these additives not only on viscous properties, but on the whole set of sealant's properties as well.

Participation of silica in structuring changes viscosity and thixotropic properties and could also exert orientation effect on PSO during curing and thus "strengthen" a formulation as filler [9, 10].

Curing of sulfur vulcanized rubbers is known to be inhibited by silica due to strong adsorption of vulcanizing group components on its surface. In addition, acidic compounds are known to inhibit interaction of sulfhy dryl groups with metal dioxides, while basic compounds act as promoters [2]. Both these facts may take place when silica is added to PSO sealants, although the deepest influence is caused by acidic properties of sealants.

Addition of acidic aerosils A-300 and A-175 (pH=3.6  4.3) [11, 12], inhibits curing process. A considerable increase of viability is observed after addition of 1 mass part of aerosil already (Fig. 5.9). The most intense increase of viability is observed after addition of A-300 aerosil because of the increase of A-300 aerosil-PSO surface contact, comparing A-175 aerosil. Addition of BS-120 (pH=8 9) changes viability of sealants in a small way independently on PSO type.

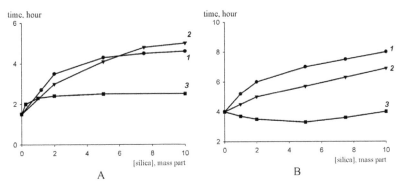

**FIGURE 5.9**  The dependence of viability of sealants based on liquid thiokol (A) and TPM-2 polymer (B) on silica content: 1 – A-300; 2 – A-175; 3 – BS-120.

Formulations with liquid thiokol demonstrate the same behavior as formulations with TPM-2 at addition of aerosils (A-300, A-175) (Fig 5.9, b). However, silica exerts deeper influence on viability t of sealants based on liquid thiokol if $MnO_2$ is used as a vulcanizing agent. It is related to higher sensitivity of liquid thiokol and its blends to variation of pH and may be explained by differences in the main chain structure of these oligomers and first of all by disulfide bonds in thiokol's main chain, which can participate in curing entering into thiol-disulfide exchange reactions.

Water exerts the most profound influence on viability among all studied glycols (Fig 5.10). Even the slightest water additives (1 mass part) result in dramatic decrease of viability of PSO-based sealants.

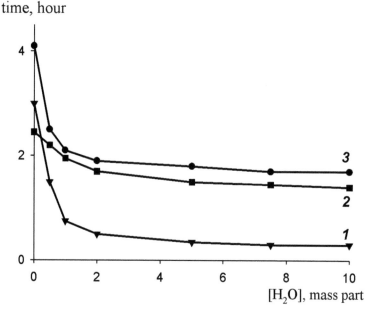

**FIGURE 5.10**  The dependence of viability of liquid thiokol sealants on the content of water: 1–A-175 = 5; 2–A-175 = 2; 3–BS-120 = 5.0 mass parts.

Its maximum drop is registered for water content of three mass parts. Water is formed in a sealant during curing in oxidation of end SH-groups by metal dioxide:

$$2 \sim RSH + MeO_2 \longrightarrow \sim R\text{-}S\text{-}S\text{-}R \sim +MeO + H_2O$$

Therefore, curing paste water is a catalyst (initiator) of oxidation process, as it shifts reaction to the right side and accelerates curing at the initial moment [4, 13].

The effect of curing promotion by the addition of water is stronger in aerosil-filled sealants (Fig. 5.10). Formulations with TPM-2 polymer demonstrate the same behavior as sealants with liquid thiokol. Further increase of water content does not change the viability of sealants. A stronger effect of water on the viability of sealants with liquid thiokol, than on formulations with TPM-2 polymer may be related to not only the activation of SH-groups but to thiol-disulfide exchange reaction as well.

EG, among other glycols, reduces slightly the viability of sealants with TPM-2 polymer. DEG and glycerol, as well as EG for thiokol do not exert a considerable influence on a curing rate.

Silica additives exert a substantial effect on the strength of sealants [8, 12, 14]. A distinct increase of strength is observed for 1 mass part of added aerosil (Fig. 5.11). Further increase of silica content in sealant leads to monotonous increase of strength. The effect of strength increase is stronger for formulations with TPM-2 polymer, than for sealants with liquid thiokol. The improvement of strength properties can be explained by the enhancement effect. In other words, silica acts as "strengthening" fillers in PSO-based sealants and the degree of strengthening correlates with the dispersity of silica.

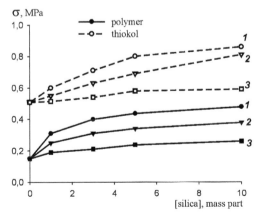

**FIGURE 5.11**    The dependence s of sealants with TPM-2 polymer (—) and liquid thiokol (– – –) on the content of silica: 1 – A-300; 2 – A-175; 3 – BS-120.

A certain increase of strength of PSO-based sealants is observed after addition of water or EG (Fig. 5.12). The effect of strength increase is known to appear for rubbers filled with light fillers and addition of such glycols as EG and triethanolamine [10]. It is explained not so much by a plasticizing effect as by sulfur vulcanization processes. In addition, glycols show properties of surfactants (dispersing agents) in the presence of light fillers, and favor their better distribution in a binding material. The increase of strength of PSO formulations may be related to formation of coagulation structures by glycols with silica. Formulations based on liquid thiokol show the increase of strength (Fig 5.12) after addition of water. However, the effect of strength increase is weaker, than for sealants with TPM-2 polymer. However, addition of large quantities of EG or water (50 mass parts and more) to TPM-2 polymer compositions does not lead to a dramatic decrease of strength. It indicates that these compounds may help in making if usable sealants.

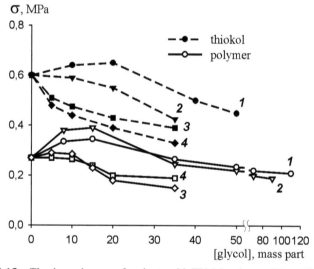

**FIGURE 5.12**   The dependence s of sealants with TPM-2 polymer (¾) and liquid thiokol (– – –) on the content of glycols (A-175 = 2 mass parts): 1 – water, 2 – EG, 3 – DEG, 4 – glycerol.

Further addition of glycols to PSO sealants lead to a plasticizing effect. This effect is the strongest for water and glycerol. DEG and glycerol act as plasticizers and reduce strength. Similar effects can be observed if A-175 aerosil is changed to HA A-300 aerosil or carbon white BS-120.

Silica exerts a profound influence on relative elongation at rupture (deformability), especially for compositions with TPM-2 polymer. The growth of relative elongation is observed for such sealants at small additives of silica (up to 5 mass parts) (Fig. 5.13). Deformability depends as a rule on the dispersity of silica: the less dispersed is additive, the higher is relative elongation. It is related to the increase of adsorption interactions and formation of hydrogen bonds by more dispersed silica. It leads, on the one hand, to the increase of strength, while, on the other hand, it reduces elastic properties.

Sealants with liquid thiokol show different behavior (Fig 5.13). Addition of silica decreases relative elongation immediately. Low-active additives (such as BS-120) exert less influence on this parameter.

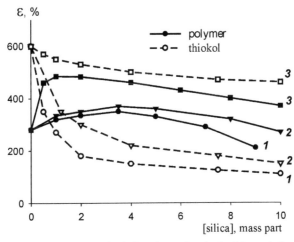

**FIGURE 5.13**   The dependence of relative elongation ($e_{rel}$) of formulations with TPM-2 polymer (—) and liquid thiokol (– – –) on silica content: 1 – A-300; 2 – A-175; 3 – BS-120.

Introduction of glycols increases relative elongation considerably (Fig. 5.14.). Glycols are ineffective as thixotropic additives, but they are the most active promoters of deformability of PSO sealants. A substantial increase of this parameter was observed after introduction of 5–10 mass parts of DEG and EG into formulations with liquid thiokol (Fig. 5.14, A). Water and glycerol are ineffective here.

As for substances with TPM-2 polymer (Fig. 5.14, b), DEG is the most suitable glycol for them. Glycerol and EG increase deformability too, but to a much less extent. Water exerts little influence on relative elongation, as for thiokol sealants. The increase of relative elongation may be explained by a plasticizing effect of DEG, which is insufficiently efficient for structuring.

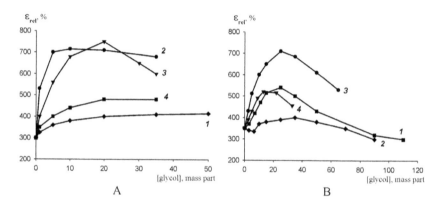

**FIGURE 5.14**    The dependence of relative elongation ($e_{rel}$) of sealants with liquid thiokol (A) and TPM-2 polymer (B)on the content of glycol (A-175 = 2 mass parts): 1 – water; 2 – EG; 3 – DEG; 4 – glycerol.

We can make a conclusion from obtained results, that glycols are not true plasticizers. Extreme variations of strength and relative elongation with the increase of glycol content can be explained by their compatibility with PSO in solubility parameter (dp).

Data from Table 5.1 demonstrate, that all glycols are thermodynamically incompatible with liquid thiokol and TPM-2 polymer. Their small additives (1–2%) seem to behave as structural plasticizers and increase strength and relative elongation. Technological "compatibility"appears at high concentrations of glycols, when DEG and glycerol act as plasticizers. Glycol content >20 mass parts decreases strength and relative elongation due to incompatibility with PSO and separation of glycols to a separate phase. DEG and EG have the best plasticizing properties, as their dp is the closest to that of PSO (Table 5.1).

**TABLE 5.1**  Solubility Parameters of Glycols and PSO [15]

| Glycol, PSO | $\delta$, $(MJ \cdot m^3)^{0.5}$ |
|---|---|
| Water | 47.7 |
| Ethylene glycol | 28.9 |
| Glycerol | 33.6 |
| Diethyleneglycol | 26.9 |
| Thiokol | 18÷19.2 |
| TPM-2 polymer | 17.6 (calculated) |

Therefore, researches have shown, that A-300 and A-175 aerosils are the most active thixotropic additives for construction sealants. Effective glycols have been found and their application limits have been set. Combination of glycols and silica has been found to lead to a dramatic increase of silica efficiency and thus its content in sealants can be reduced.

It should be noted, that carbon white is less active as thixotropic additive, but its combination with glycols is rather prospective for seam sealants.

Optimal dosages of silica and glycols have been found, considering their influence on viscosity, thixotropicity, viability, deformation and strength properties of sealing compositions. The quantity of these additives required for thixotropic construction sealants depends on the degree of filling, the nature of filler, oligomer and a thixotropic additive itself.

## 5.2  THE INFLUENCE OF FILLERS ON CURING AND PROPERTIES OF SEALANTS

Introduction of fillers obviously exerts a considerable influence on mobility of polymer's kinetic units, structuring processes (first of all, on the filler-polymer boundary), kinetics and degree of curing, structural, physic-chemical and mechanical characteristics of curable filled polymers [9, 16]. It relates to thiokol sealants either, as they are used only in the filled state, because it improves its properties, especially deformation and strength. The variation of properties of polymer compositions with the increase of filler content usually has the maximum and further deterioration of parameters. According to Yu. S., Lipatov, it is related to so called concentration optimum, which is considered to be the ultimate saturation of filler's surface adsorption centers by macromolecules. When the content of filler exceeds this optimum, network structure becomes discontinuous, its specific surface and the quantity of micro defects grow resulting in the decrease of all the range of properties [9].

The effect of viscosity increase with improvement of physic-mechanical properties also requires consideration, as it may deteriorate performance characteristics of compositions.

Such strengthening fillers as BS-50 carbon white and P–803 carbon black exert the deepest influence on properties of thiokol sealants [2, 4]. Carbon black strengthening effect is related to a developed surface (the specific surface value is $-15$ $m^2$ g), good dispersive ability and activity of surface thanks to various functional groups forming physical bonds with PSO. The influence of nature, dispersity and carbon black content on rheological and physico-chemical properties of thiokol sealants has been studied earlier [2, 17–20].

The analysis of PSO (mainly liquid thiokol) world consumption in the recent 20 years revealed, that more than 60% of produced PSO is used for production of sealants for glass packets. In addition, some PSOs (usually thiol-containing ones) are used for making of sealants designed for sealing interpanel seams in house building, especially in Russia [21–23]. Typical requirement for aforementioned sealants are color (they must be light) and rheological properties (they must be suitable for machine processing). PSO sealants filled with chalk suit these requirements. Such sealants must keep a construction sealed in operating conditions. Therefore, they are supposed to have the following properties: high elastic properties, deformability, ability to minimal accumulation of residual deformations, high adhesion to concrete, glass and duralumin under static and retarded dynamic deformations in the temperature range from $-50$ to $+50°C$ usually at the constant contact with water. It should be noted, that there are no published researches concerning influence of carbon black and chalk as fillers on viscous, technological, physic-mechanical properties and performance of sealants with TPM-2 polymer as well as the influence of chalk on the properties of thiokol sealants. There are no studies on the influence of chalk as a filler on curing processes in thiol-containing PE or thiokol sealants. Thus, a research have been carried out to study the influence of various chalk samples on curing processes, technological and physico-mechanical properties and performance of sealants based on liquid thiokol or TPM-2 polymer [8, 24–32].

As Fig. 5.15 demonstrates, addition of natural chalk to PSO leads to a substantial increase of viscosity. While liquid thiokol formulations with added chalk behave as Newtonian liquids in the all range of shear rates, the viscosity of TPM-2 polymer formulations with added chalk deviates from Newton's law at low shear rates because of formation of weak coagulation structures via interaction of TPM-2 polymer's hydroxyl groups with active groupings on chalk's surface (Fig 5.15B). Such effects are observed for hydrophobic and chemically precipitated chalks. The analysis of obtained results leads to the following conclusions:

1. The viscosity of filled compositions depends on the dosage of chalk and not on its nature.
2. Formulations with TPM – 2 polymer show a distinct thixotropic effect at low shear rates in a form of viscosity increase, which is related to the structure of a polymer itself.
3. Compositions with TPM – 2 polymer are less viscous, than liquid thiokol blends assuming filling was the same. The main reason of this effect is the viscosity caused by PSOs themselves.

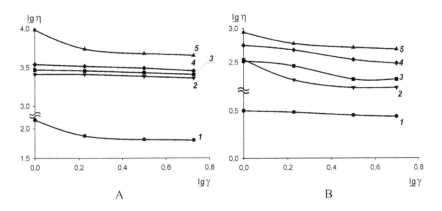

**FIGURE 5.15**   The dependence of viscosity (lgh) and shear rate (lgg) of compositions with liquid thiokol (A) and TPM-2 polymer (B) on the type and content of natural chalk: 1–no chalk, 2–60 mass parts, 3–80 mass parts, 4–100 mass parts, 5–120 mass parts.

Comparison of sealant viability (Fig 5.16) has shown that the nature of added chalk influences the rate of curing. The most intense decrease of viability occurs in the presence of chemically precipitated chalk, while hydrophobic chalk is the least effective. The more natural chalk is introduced, the more constituent catalytic additives thiokol gets, but their effect on the viability of liquid thiokol formulations is achieved only if chalk's content is above 80 mass parts. The increase of curing rate (decrease of viability) seems to be related, first of all, to the decrease of chalk particles' sizes and corresponding growth of surface contact area between PSO and filler with intensification of adsorption interactions. Indeed, the specific surface of chemically precipitated chalk is 5–20 $m^2$g depending on its production method, while its value for natural chalk is 0.6–1.0 $m^2$g [32]. A minimal activity (inhibiting effect) of hydrophobic chalk is related to the acidic surface of its particles, which is covered by a thin layer of synthetic fatty acids (SFA), because SFA are known to be effective retarders of liquid thiokol curing by $MnO_2$ [1]. Considering the fact, that the content of SFA in liquid thiokol can reach 2% mass, as well as that chalk content in sealant reaches 100–150 mass parts, SFA can be a relatively strong retarder for the process of liquid thiokol curing by $MnO_2$ that is really observed in practice. The increase of viability (the decrease of curing rate) caused by extra chalk has similar behavior and does not depend on chalk's nature (Fig. 5.16). Two reasons can stipulate this effect:

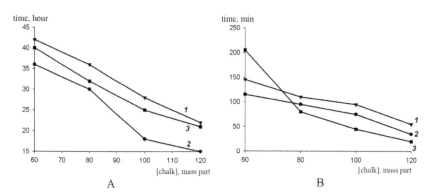

**FIGURE 5.16**   The dependence of viability of liquid thiokol (A) and TPM-2 polymer sealants (B) on the nature and content of chalk: 1–hydrophobic chalk, 2–chemically precipitated chalk, 3–natural chalk.

1.  more reactive forms of PSO could accumulate at the interface via adsorption due to orientating effect of filler's surface and emerging of favorable conditions for interaction of PSO end SH-groups with MnO2 [9, 10].

2.  the more chalk is added, the more viscous a composition becomes and Therefore, gel formation occurs. It usually correlates with the loss of viability at earlier stages of curing, especially when oligomers with functionality of 3 are used [33, 34].

Comparison of Figs 5.16a and 5.16b leads to the conclusion, that TPM − 2 polymer is less active in $MnO_2$ oxidation reactions, than liquid thiokol. It may be related to different structure and viscosity of PSO as well as by inevitable presence of catalytic impurities (such as iron salts) [2, 35] and sulfur in PSO blends. The latter one is known to participate in curing reactions like this:

$$2\sim RSH + S \longrightarrow \sim R\text{-}S\text{-}S\text{-}R\sim + H_2S\uparrow$$

and to be a highly effective initiator [2, 19]. Preparation of TPM-2 polymer sealants with viability similar to that of liquid thiokol sealants requires implication of a more active curing system. Both this requirement and the results of our studies have been considered in practice, in particular, at production of SG–1 mastic compound.

The increase of chalk content leads in all cases to the increase of strength at rupture values for cured formulations (see Fig. 5.17, a, b), while the strength of thiokol formulations is higher, than the strength of TPM –2 polymer formulations.

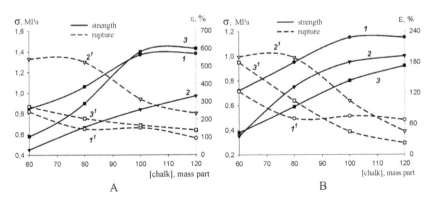

FIGURE 5.17    The dependence of strength at rupture (1, 2, 3) and relative elongation (1', 2', 3') on the nature and content of chalk for sealants based on liquid thiokol (A) and TPM-2 polymer (B): 1,1'–chemically precipitated chalk; 2,2'–hydrophobic chalk; 3,3'–natural chalk.

Chalk has the following strengthening effect on systems based on liquid thiokol depending on nature: finely dispersed chemically precipitated chalk > natural chalk > hydrophobic chalk, while for TPM-2 polymer systems: chemically precipitated hydrophobic chalk > natural chalk. The most intense strengthening is observed for 80–100 mass parts of added chalk independent on its nature. Differences in strengthening effect can be first of all explained by adsorption activity of chalk's surface to PSO. Strengthening effect seems to be directly related to intensification of interaction at PSO-chalk interface. The less particles are and the better filler distributes in sealant, the better filer-thiokol contact is. It favors the intensification of physical interactions and must increase the effective density of transversal networks ($n_{eff}$). A slight intensification effect caused by hydrophobic chalk can be explained by inhibition of oligomer-filler physical interactions by the modifier (SFA) adsorbed on chalk surface. Low activity of natural chalk seems to be primarily related to a small specific surface. It should be noted, that the strengthening effect of fillers is stronger for TPM-2 polymer sealants, than for blends with liquid thiokol (Fig. 5.18).

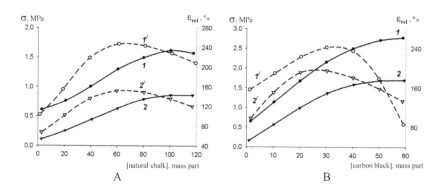

**FIGURE 5.18** The influence of PSO nature, the content and nature of filler on the strength at rupture (1,2) and relative elongation values (1', 2') (A-natural chalk, B–carbon black): 1, 1'–liquid thiokol, 2,2'–TPM-2 polymer.

Indeed, addition of fillers increases the strength of sealants thiokol sealants 2–2, 5 times (for chalk) and 3–4 times for carbon black, while sealants with TPM-2 polymer demonstrate 4–6 times increase for chalk and 7–8 times increase for carbon back. It is evidently due to the different in structure of oligomers. Fillers may exert deeper orienting influence on the chain of TPM-2 polymer, than of thiokol because of stronger intermolecular interactions [8–10]. Similar effects of the same nature are observed in elastomeric rubbers [36].

Information obtained from physic-mechanical properties correlates with estimations of transversal bond density made for sealants (Fig 5.19). The increase of chalk content always leads to the growth of transversal bond density in liquid thiokol systems. Its maximum is reached, when the content of natural or chemically precipitated chalk is 100 mass parts per 100 mass parts of PSO. Observed decrease of total transversal bond density in liquid thiokol formulations with more than 100 mass parts of fillers is evidently related to the increase of defectiveness of forming three-dimensional network due to formation of big branched molecules with less mobility. It results in limited ability of reactive groups to enter into curing reaction and participate in formation of physical network due to"undercured" layer of liquid thiokol being on filler's surface [36–38]. The chem value for chalk filled sealants is lower, than for carbon black filled sealants [8, 32]. The effective density of transversal bond network for TPM-2 polymer sealants is, as expected, 10–50% lower, than for liquid thiokol sealants. It reaches maximum, when 100–120 mass parts of chalk are added [32].

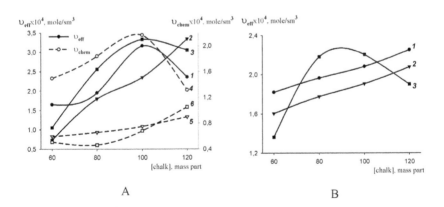

**FIGURE 5.19**   The influence of chalk's nature and content on n value of thiokol (A) and TPM-2 polymer sealants (B): 1,4–chemically precipitated chalk; 2,5–hydrophobic chalk; 3,6–natural chalk. 1,2,3–$n_{eff}$. 4,5,6–$n_{chem}$.

Less strengthening effect of added chalk can be first of all stipulated by the absence of chain-like structures forming at particular content of carbon black (~30 mass parts), and also by small specific surface areas as well as by small content of polar reactive groups on the surface of chalk particles able to participate in adsorption interactions with PSO. Observed increase of strength may be stipulated by the activating influence of chalk surface on curing processes due to oligomer's adsorption and emerging conditions favoring better conversion in the reaction of SH-groups with manganese dioxide. It may be the reason of increasing strength properties of TPM-2 polymer sealants considering slight growth (from 10% to 30%) of total transversal bond density (t.b.d.). Higher values of effective t.b.d. for TPM-2 formulations, especially for small chalk concentrations, seem to be caused by participation of constituent OH-groups in adsorption interaction with filler.

The difference in strengthening effects caused by various types of chalk can be first of all explained by the difference in dispersity (specific surface) of fillers and filler-PSO contact surface correspondingly, considering small chemical activity of chalk's surface.

The main reason of changing curing conditions of reactive oligomers in the presence of filler is their adsorption interaction with a surface having an impact on both process kinetics and final properties of cured oligomer [38]. Forming adsorptive layer of macromolecules has limited range of conformations and is much less mobile, that may restrict reactivity of functional groups in macromolecules. Therefore, curing process may retard, that is really observed for some systems [38–40]. Such fillers as carbon black, titanium dioxide, silicon dioxide are catalysts for curing process thanks to a lot of functional groups and metal containing impurities on their surface, which are able to accelerate curing process. It may be the explanation for acceleration of curing process in the presence of carbon black [18]. So, liquid Thiokol's curing process has been determined to accelerate in the presence of carbon black independently on curing agent type due to adsorption of oligomers on carbon black surface and catalysis of PSO's SH-groups oxidation by various functional groups and impurities situated on this surface. The increase of carbon black dispersity causes considerable impairment of its distribution in liquid thiokol. Although Thiokol's molecular mobility changes in this case, strength properties of sealant deteriorate and total t.b.d. value reduces. A slight change of NMR adsorption line width with the increase of filler content indicates, that liquid Thiokol carbon black interaction is mainly physical [20]. Curing of reactive oligomers is usually carried out faster with fillers, than without it, due to orienting effect of filler on macromolecules and formation of their more reactive forms in adsorption layer [36, 41–42]. In addition, promoting effect of filler ought to increase with dispersity because surface contact increases and more oligomer macromolecules can be found in adsorption layer. However, formation of three-dimensional structures in the presence of filler may increase defectiveness of forming structures [37].

$MnO_2$ of two types was used as a curing agent. Its excess factor (molar) to SH-group was 3. The curing agent of the first type has been determined to be 40 times more active, than the second one in curing rate and viability for NVB–2 liquid Thiokol.

Curing process rate was monitored by the change of spin-spin relaxation time $T_2$ in proton spic echo experiments (Carr-Parcell method), NMR frequency was 5, 6 MHz and the temperature was 200C [26]. Observation time was one day.

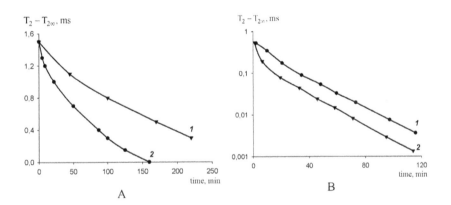

**FIGURE 5.20**  Kinetic dependences for curing of sealants with TPM-2 polymer (A) and liquid thiokol (B) by active $MnO_2$: 1–no chalk; 2–50% (weight) of chalk.

Figure 5.20 represents semi-logarithmic anamorphisms of kinetic curves $(T_2-T_{2\infty})$–t (TPM-2 polymer's curing time) (Fig. 5.20A) and liquid thiokol (Fig 5.20B) with 0 and 50% weight of chalk, cured by active $MnO_2$, where t – curing time, $T_{2\infty}$–spin echo in cured samples, which characterizes the maximum degree of conversion of oligomeric functional groups. $T_{2\infty}$ values of cured sealants are given in Table 5.2. depending on chalk content, as well as curing rate values K, which the reciprocal value of the time $T_2$ decays e times.

Table 5.2 demonstrates, that both values are in symbiosis with chalk content. Formulations with TPM-2 polymer cure considerably (2–2, 5 times) slower, than, formulations with NVB–2 [30].

**TABLE 5.2**   The maximum $T_2$ value ($T_{2\infty}$) and the curing rate k for PSO curing by active manganese dioxide depending on chalk content

| Chalk content, % (weight) NVB–2 | $T_{2\infty}$, ms | | $k \times 10^2$, min$^{-1}$ | |
|---|---|---|---|---|
| | | TPM-2–polymer | NVB–2 | TPM-2-polymer |
| 0 | 0.425 | | 0.789 | 6.62 | 2.62 |
| 9 | 0.444 | | 0.776 | 9.17 | 4.71 |
| 25 | 0.449 | | 0.793 | 7.87 | 3.57 |
| 40 | 0.453 | | 0.842 | 7.81 | 5.32 |
| 50 | 0.461 | | 0.954 | 12.0 | 6.17 |

It may be stipulated by several reasons: different structure, molecular weight or functionality. However, it seems to be first of all stipulated by the lack of so called "free" sulfur in TPM-2 polymer formulations, which is always present in small concentrations (0,1–0,3 mass parts) in liquid thiokol formulations.

Results [27–30] from Table 5.2 and Fig. 5.20 lead to a conclusion, that addition of chalk has more promoting effect on curing processes in TPM-2 polymer formulations, than in formulations with liquid thiokol. High curing rate may cause negative consequences for the completeness of curing process. Indeed, as T values show (Table 5.2), the completeness of curing decreases in PSO formulations with added chalk, especially for TPM-2 polymer.

Figure 5.21 and Table 5.3 contain similar data for compositions with NVB-2 thiokol, cured by low-active $MnO_2$.

**TABLE 5.3**   Maximum values $T_{2\infty}$ and curing rate k for NVB-2 thiokol curing by low-active manganese dioxide

| Chalk content, % weight | $T_{2\infty}$, ms | $k \bullet 102$, min$^{-1}$ |
|---|---|---|
| 0 | 0.423 | 0.508 |
| 40 | 0.510 | 0.806 |
| 50 | 0.604 | 2.78 |

It is interesting, that filler has stronger influence on sealants with low curing rate. However, $T_{2\infty}$ value does not depend on the activity of a curing agent for unfilled formulations with liquid thiokol (Tables 5.2 and 5.3).

An evident suggestion here is that if sealant's curing rate reduces, a boundary layer cannot achieve equilibrium state, which has better conditions for reaction because of aforementioned reasons. It seems to explain the more substantial effect of filler if a low-active $MnO_2$ is used.

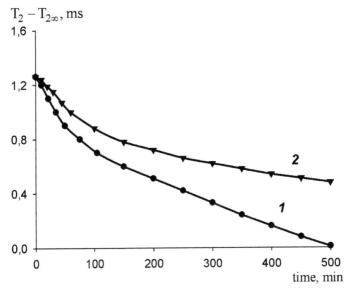

**FIGURE 5.21**   Kinetic curves for NVB-2 thiokol curing by low-active manganese dioxide: 1-no chalk; 2–50% (weight) of chalk.

Another important conclusion, that arises from comparison of Tables 5.2 and 5.3 and Fig. 5.22, is that chalk-filled PSO compositions cure faster, than unfilled ones, and this effect is independent on PSO type.

The increase of PSO curing rate is caused by all types of chalk and is higher for TPM-2 polymer formulations (Fig. 5.22). The observed activity of chalks correlates with their influence on strength properties of sealants and on t.b.d. [32].

The increase of curing rate observed with addition of filler (Fig. 5.22) is stipulated by more rapid formation of chemical mesh points, which is, in turn, in symbiosis with consumption of functional groups. Therefore, despite the decrease of molecular mobility near a filler surface layer, reaction rate is higher I the presence of filler, than in uncured system.

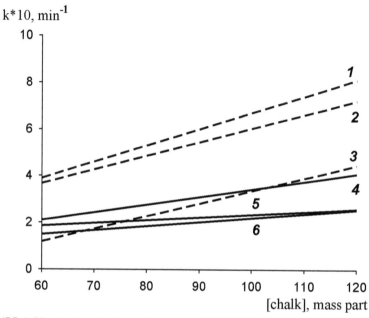

**FIGURE 5.22**   Curing kinetics of sealants with TPM-2 polymer (1, 2, 3) and liquid thiokol (4,5,6) depending on the nature and content of chalk: 1,4–chemically precipitated chalk; 2,5–hydrophobic chalk; 3,6–natural chalk.

We can suppose that the increase of curing rate in the presence of filler is caused by accumulation of more reactive forms of interacting macromolecules (their functional groups) in a boundary layer due to adsorption processes. It may be stipulated by the increase of ordering of reacting macromolecules (it promotes reaction, according to Ref. [10]).

A promoting effect may be also caused by the activity of OH–groups, which can be present on a surface of many mineral fillers, including chalk, as well as by the presence of water in it.

The decrease of maximum curing degree at the presence of filler and related $T_{2\infty}$ growth can be the effect of forming physical network consisting of macromolecules and filler particles [42], where latter ones act as poly functional transversal bonds. Therefore, conversion of functional oligomeric groups reduces, in the gel-formation point [38]. Gel formation (and related restrictions to the mobility of macromolecules) on earlier curing stage will certainly impede reaction and Therefore, decrease the transversal bond density, as it is observed for PSO sealants with added chemically precipitated chalk, which is more active type of chalk (Fig. 5.19). $T_{2\infty}$ value characterizes mobility of cured samples and can be an indicator for this process. Indeed, data in Table 5.2 show, that $T_{2\infty}$ in symbiosis with chalk content for TPM-2 polymer. This tendency is weaker for liquid thiokol (Table 5.2), however it reveals nicely, when low-active $MnO_2$ is used.

In a whole, chalk additives to liquid thiokol sealants increase both chemical and total density of transversal bonds, but to a lesser extent, than carbon black [28, 32]. A "strengthening effect" that is the increase of strength of thiokol sealants, is Therefore, weaker.

Thus, impulse NMR studies have found that:
1. natural chalk additives promote PSO curing $MnO_2$, a stronger effect is observed for formulations with TPM-2 polymer;
2. chalk influence on curing processes is deeper, when low-active $MnO_2$ is used;
3. chalk activity mainly depends on its dispersity (the area of contact with PSO). The highest curing rate is thus observed for chemically precipitated chalk;
4. chalk additives increase both physical and chemical density of transversal bonds that can be explained by orienting effect of chalk surface on PSO macromolecules.

Summarizing information on the influence of fillers on the properties of PSO sealants, we can conclude, that chalk-PSO and carbon-black-PSO interactions fall into the third group of Berlin's classification [41], considering observed effects of strengthening, the increase of physical network density and changing of other properties of sealing compositions. Adhesion parameter ($x$) (introduced here as Sato and Furukawa concept [42]) of such unfilled compositions is in the range of $0 < x < 1$, when PSO adsorbs on filler's surface and they can physically interact.

## 5.3   THE INFLUENCE OF FILLERS ON CURING AND PROPERTIES OF THIOURETHANE SEALANTS

The influence of fillers on curing kinetic and properties of PSO sealants is well described in publications by Averco-Antonovich, L. A., Minkin, V.

S. and Nefed'ev, V. S. with coauthors [1, 8, 17–20, 25–31]. On the other hand, the influence of fillers on curing and properties of thiourethane sealants is not studied well, except [43, 44].

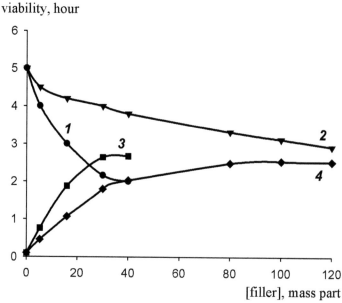

viability, hour

**FIGURE 5.23** The dependence of viability of thiourethane sealants based on SKUPFL–100 (0,04 mass parts of OM–3 catalyst) on the degree of filling: 1 – thiokol, carbon black; 2 – thiokol, chalk; 3 – TPM-2 polymer, carbon black; 4 – TPM-2 polymer, chalk.

The dependence of viability of thiourethane compositions on the content of curing agents is discussed for thiourethane sealants based on SKUPFL–100 and represented in Fig. 5.23 (formulations based on SKUP-PL–5003 and polyisocyanate behave in a similar way). Variation of viability with addition of filler depends on PSO nature [4, 46]. Thiourethane sealants with thiokol demonstrate effects similar to thiokol sealants, i.e., the increase of filler content causes the decrease of viability [31, 45]. It is explained by adsorption phenomena arising with the increase of filler content in compositions. They increase the viscosity of medium, favor longer lifetimes of labile bonds between SH– and NCO–groups [47] and

Therefore, accelerate curing. It is confirmed by curing kinetics estimations made for cured formulations with Thiokol and SKUPFL–100 using NMR (Fig. 5.24).

When filler content increases, curing occurs faster. The lowest $T_2$ value at the initial stage (i.e., the maximum degree of curing) is shown by formulations with maximum filler content (Fig 5.24, A).

Carbon black is better curing promoter, than chalk, because of more active and developed surface and ability to catalyze curing processes [19].

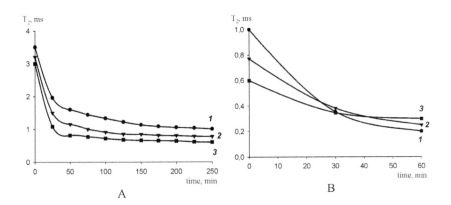

**FIGURE 5.24**    $T_2$ variation with the time of thiourethane sealant curing (thiokol + SKUPFL–100, equimolar ratio) (A – the initial stage; B – the final curing stage ("0" point on B corresponds to 250 min)) on the content of carbon black: 1–15 mass parts; 2–30 mass parts; 3–45 mass parts.

TPM-2 polymer sealants demonstrate opposite effect. It is obviously thanks to the presence of both hydroxyl and SH-groups in TPM-2 polymer and their possible mutual activation [48]. Fillers added to thiourethane sealants seem to retard this mutual activation. The curing rate of thiourethane sealants based on TPM-2 polymer reduces with the increase of carbon black content that is confirmed by NMR and IR data (Fig. 5.25A and 5.26.).

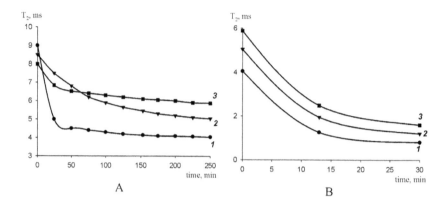

**FIGURE 5.25** $T_2$ variation with the time sealant curing (TPM-2 + SKUPFL–100, equimolar ratio) (A – the initial stage; B – the final curing stage ("0" point on B corresponds to 250 min)) on the content of carbon black: 1 –15 mass parts; 2–30 mass parts; 3–45 mass parts.

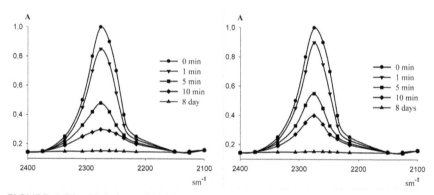

**FIGURE 5.26** Variation of NCO-group signal in thiourethane sealant with TPM-2 polymer and PIC during curing (A – no filler; B – 5 mass parts of chalk) at different times

IR spectroscopy studies have been carried out for sealants with no filler and for compositions with 5 mass parts of chalk (per 100 mass parts of oligomers), because spectra with good quality cannot be obtained from samples with more chalk.

Considering slower consumption of NCO-groups in filled compositions, it is obvious, that it exerts a negative influence on the rate and conversion at the first stage of curing of thiourethane sealants based on TPM-2.

Carbon black is much more active curing retarder for thiourethane sealants based on TPM-2 polymer, than chalk, that can also be caused by its more developed and active surface.

Fillers exert negative influence on the final stage of curing in all studied cases (see Figs. 5.24B and B). The influence of fillers on the degree of curing is, obviously, the same and independent on types of PSO and isocyanate-containing component. Its effect is partial undercure of oligomers (PSO in this case) in the presence of solid phase due to limited mobility of macromolecules, especially for high-filled formulations. It is the reason of some $T_2$ increase. Same effects are demonstrated by thiokol sealants [31].

The properties of carbon black filled sealants with thiokol and prepolymers (SKUPFL–100 and SKUPPL–5003) are represented in Table 5.4.

As we can see, 30 mass parts of carbon black is optimal additive for strength properties indicating formation of continuous filler phase in bulk sealant. The increase of carbon black content above 30 mass parts deteriorates elastic properties of a sealant. The hardness is proportional to a carbon black quantity. The decrease of adhesion to concrete with more added filler is caused by suppression of diffusion processes on a "sealant-concrete" interface boundary and worse conditions for of adhesion contact.

Thiokol sealants with added chalk show somewhat another behavior (Table 5.5). The relative elongation of formulations decreases with more added chalk. Although considered as a non-strengthening filler, chalk has some strengthening effect on thiourethane sealants, but it is weaker, than carbon black.

Adhesion to concrete increases with chalk content in sealant. Despite sealants filled with carbon black, chemical affinity of concrete (the binding agent is cement (consists of silicate $CaSiO_3$ and calcium aluminates $CaO * Al_2O_3$) and chalk ($CaCO_3$) may be the reason. Therefore, adhesion of chalk filled thiourethane sealants to concrete continuous to grow even when the viscosity is high (corresponds to the viscosity of the formulation with 40 mass parts of carbon black). The behavior of strength and hardness properties is similar to those of carbon black filled formulations, although these properties are much better for formulations filled by carbon black, than for formulations filled by chalk.

**TABLE 5.4** The influence of carbon black content on the properties of sealants with thiokol and urethane prepolymers (equimolar ratio of components; 0.01 mass parts of OM–3 catalyst)

| Carbon black, mass parts | $s_{conv}$, MPa | $e_{rel}$, % | $A_{con}$, MPa | H, units | $Q_{water}$, % | $Q_{toluene}$, % | $n_{eff}$•$10^4$, mole'sm$^3$ | $n_{chem}$•$10^4$, mole'sm$^3$ |
|---|---|---|---|---|---|---|---|---|
| Thiokol and SKUPFL–100 | | | | | | | | |
| 0 | 1.0 | 250 | 1 coh. | 20 | 17.3 | 134 | 2.2 | 0.9 |
| 10 | 2.6 | 270 | 1.3 adh. | 35 | 14.2 | 123 | 2.7 | 1.1 |
| 20 | 3.8 | 285 | 1.6 adh. | 50 | 12.8 | 112 | 3.3 | 1.3 |
| 30 | 5.2 | 300 | 1.8 adh. | 60 | 11.4 | 95 | 3.8 | 1.5 |
| 40 | 5.4 | 190 | 1.6 adh. | 70 | 10.1 | 87 | 4.3 | 1.7 |
| Thiokol and SKUPPL–5003 | | | | | | | | |
| 0 | 0.6 | 100 | 0.6 coh. | 10 | 15.8 | 193 | 1.7 | 0.4 |
| 10 | 1.6 | 145 | 0.8 adh. | 25 | 12.3 | 171 | 2.4 | 0.6 |
| 20 | 2.3 | 180 | 1.0 adh. | 35 | 10.6 | 154 | 2.9 | 0.8 |
| 30 | 3.1 | 220 | 1.2 adh. | 50 | 9.8 | 145 | 3.5 | 0.9 |
| 40 | 3.3 | 140 | 1.0 adh. | 60 | 8.7 | 140 | 3.8 | 1.0 |

**TABLE 5.5** The influence of natural chalk content on the properties of thiourethane sealant based on thiokol (equimolar ratio of components; 0.01 mass parts of OM-3 catalyst)

| Carbon black, mass parts | $s_{conv}$, MPa | $e_{rel}$, % | $A_{con}$, MPa | H, units | $Q_{water}$, % | $Q_{tolu\text{-}ene}$, % | $n_{eff}$•$10^4$, mole'sm$^3$ | $n_{chem}$•$10^4$, mole'sm$^3$ |
|---|---|---|---|---|---|---|---|---|
| Thiokol and SKUPFL–100 | | | | | | | | |
| 0 | 1.0 | 250 | 1 coh. | 20 | 17.3 | 134 | 2.2 | 0.9 |
| 40 | 1.3 | 250 | 1.3 coh. | 30 | 14.5 | 121 | 3.3 | 0.9 |
| 80 | 1.7 | 230 | 1.6 adh. | 50 | 12.3 | 104 | 3.8 | 1.0 |
| 120 | 2.1 | 210 | 1.7 adh. | 60 | 10.7 | 89 | 4.2 | 1.0 |
| 160 | 2.2 | 200 | 1.8 adh. | 70 | 9.6 | 78 | 4.5 | 1.1 |
| Thiokol and SKUPPL–5003 | | | | | | | | |
| 0 | 0.6 | 100 | 0.6 coh. | 10 | 15.8 | 193 | 1.7 | 0.4 |
| 40 | 0.8 | 110 | 0.8 coh. | 25 | 11.2 | 165 | 2.5 | 0.4 |
| 80 | 1.2 | 120 | 1.1 adh. | 40 | 9.5 | 151 | 3.0 | 0.5 |
| 120 | 1.5 | 100 | 1.3 adh. | 50 | 8.7 | 138 | 3.5 | 0.5 |
| 160 | 1.7 | 80 | 1.4 adh. | 60 | 8.5 | 124 | 3.9 | 0.6 |

Fillers have much stronger strengthening effect on thiourethane sealants with TPM-2 polymer, than on thiokol formulations (Tables 5.6 and 5.7). It is obviously due to lower polarity of TPM-2 polymer's main chain and related growth of filler's influence on properties of such compositions (the strengthening effect of fillers is known [9, 49] to be weaker in polar polymers, such as polyurethanes). The influence of fillers on sealant adhesion to concrete is also similar to thiokol-based compositions: carbon black additives above the critical value deteriorate this property. However, this effect is not observed for chalk. The critical content of filler leading to deterioration of deformation and adhesive properties (when carbon black is used for filling) as well as strength growth retardation of thiourethane sealants with TPM-2 polymer is ~40 mass parts of carbon black, ~120 mass parts of chalk. It is more, than a corresponding value for thiourethane sealants with thiokol (~30 mass parts of carbon black, ~80 mass parts of chalk). It can be explained by lower viscosity of TPM-2 polymer, than of thiokol.

**TABLE 5.6** The influence of carbon black content on the properties of thiourethane sealant based on TPM-2 polymer (equimolar ratio of components; 0.01 mass parts of OM–3 catalyst)

| Carbon black, mass parts | $s_{conv}$, MPa | $e_{rel}$, % | $A_{con}$, MPa | H, units | $Q_{water}$, % | $Q_{toluene}$, % | $n_{eff} \cdot 10^4$, mole$\cdot$sm$^3$ | $n_{chem} \cdot 10^4$, mole$\cdot$sm$^3$ |
|---|---|---|---|---|---|---|---|---|
| TPM-2 polymer and SKUPFL–100 | | | | | | | | |
| 0 | 0.5 | 160 | 0.5 coh. | 15 | 19.4 | - | 1.4 | - |
| 10 | 1.2 | 180 | 1.2 coh. | 20 | 16.7 | - | 2.0 | - |
| 20 | 1.6 | 210 | 1.5 adh. | 30 | 14.9 | 285 | 2.6 | 0.5 |
| 30 | 2.8 | 220 | 1.9 adh. | 40 | 13.4 | 240 | 3.1 | 0.8 |
| 40 | 3.7 | 240 | 2.1 adh. | 45 | 12.1 | 215 | 3.4 | 1.1 |
| 60 | 3.9 | 100 | 1.5 adh. | 60 | 10.3 | 183 | 3.7 | 1.3 |

**TABLE 5.6** *(Continued)*

| Carbon black, mass parts | $s_{conv}$, MPa | $e_{rel}$, % | $A_{con}$, MPa | H, units | $Q_{water}$, % | $Q_{toluene}$, % | $n_{eff} \cdot 10^4$, mole·sm³ | $n_{chem} \cdot 10^4$, mole·sm³ |
|---|---|---|---|---|---|---|---|---|
| TPM-2 polymer and SKUPPL-5003 | | | | | | | | |
| 0 | 0.2 | 80 | 0.2 coh. | 10 | 17.2 | 114 | 3.2 | 1.2 |
| 10 | 1.1 | 110 | 0.9 adh. | 20 | 15.6 | 88 | 3.7 | 1.6 |
| 20 | 1.9 | 130 | 1.2 adh. | 30 | 14.0 | 72 | 4.2 | 1.9 |
| 30 | 2.7 | 140 | 1.5 adh. | 40 | 11.8 | 60 | 4.6 | 2.2 |
| 40 | 3.5 | 150 | 1.8 adh. | 50 | 10.4 | 53 | 5.0 | 2.4 |
| 60 | 4.0 | 110 | 1.6 adh. | 60 | 9.5 | 45 | 5.3 | 2.6 |

**TABLE 5.7** The influence of natural chalk content on the properties of thiourethane sealant based on TPM-2 polymer (equimolar ratio of components; 0,01 mass parts of OM–3 catalyst)

| Carbon black, mass parts | $s_{conv}$, MPa | $e_{rel}$, % | $A_{con}$, MPa | H, units | $Q_{water}$, % | $Q_{toluene}$, % | $n_{eff} \cdot 10^4$, mole·sm³ | $n_{chem} \cdot 10^4$, mole·sm³ |
|---|---|---|---|---|---|---|---|---|
| TPM-2 polymer and SKUPFL–100 | | | | | | | | |
| 0 | 0.5 | 160 | 0.5 coh. | 15 | 19.4 | - | 1.4 | - |
| 40 | 1.0 | 180 | 1.0 coh. | 30 | 15.6 | 274 | 1.9 | 0.3 |
| 80 | 1.5 | 200 | 1.5 coh. | 50 | 12.3 | 242 | 2.4 | 0.3 |
| 120 | 1.8 | 160 | 1.8 coh. | 60 | 10.5 | 213 | 2.9 | 0.4 |
| 160 | 2.2 | 130 | 2.2 coh. | 65 | 9.2 | 181 | 3.4 | 0.4 |
| TPM-2 polymer and SKUPPL–5003 | | | | | | | | |
| 0 | 0.2 | 80 | 0.2 coh. | 10 | 17.2 | 114 | 3.2 | 1.2 |
| 40 | 0.5 | 100 | 0.5 coh. | 25 | 13.1 | 81 | 3.8 | 1.3 |

**TABLE 5.7**  *(Continued)*

| Carbon black, mass parts | $S_{conv}$, MPa | $e_{rel}$, % | $A_{con}$, MPa | H, units | $Q_{water}$, % | $Q_{toluene}$, % | $n_{eff} \cdot 10^4$, mole'sm³ | $n_{chem} \cdot 10^4$, mole'sm³ |
|---|---|---|---|---|---|---|---|---|
| 80 | 0.9 | 120 | 0.8 adh. | 45 | 11.0 | 59 | 4.4 | 1.3 |
| 120 | 1.5 | 120 | 1.4 adh. | 55 | 9.5 | 46 | 5.0 | 1.4 |
| 160 | 1.8 | 90 | 1.8 coh. | 60 | 8.8 | 37 | 5.5 | 1.5 |

Tables 5.4–5.7 demonstrate that fillers exert the strongest influence on systems with good compatibility of components, assuming they contain PSOs of the same type. Fillers exert maximum influence on {TPM-2 polymer + SKUPPL–5003} formulation off all studied compositions, as its components are absolutely compatible and are the least polar.

The degree of strengthening of thiourethane sealants is also determined by the concentration of forming thiourethane (urethane) bonds. The more this concentration is, the more the strengthening effect is Ref. [49]. It has not been revealed by the comparison of studied systems, because compatibility and polarity factors are considered to be more important. However the analysis of the influence of fillers on the properties of {thiokol + SKUPFL–100} formulations (i.e. the blend of well-compatible polar oligomers) and a "classical" thiokol sealant has revealed, that sealants influence more on the latter one. The explanation is the decrease of filler's strengthening effect caused by polar thiourethane bonds in thiourethane sealants.

The degree of swelling in various media (water, toluene etc.) decreases for all studied formulations with the increase of filler content in a sealant and has a strong dependence on the nature of components of thiourethane sealants and especially on their compatibility. Highly filled blends are hard to mix due to their high viscosity, whichmay make a formulation heterogeneous and Therefore, increase swelling degree of cured sealants.

Availability of raw material resources and a relative cheapness of chalk make it more preferable for production of construction sealants, than carbon black. Considering strength requirements to the strength of seam sealants (at least 0.1 MPa) [126] as well as for formulations designed to seal airfield seams (1.0–1.5 MPa) [1, 50], chalk-filled thiourethane sealants have been designed with the reserve of this parameter. A required level of relative elongation can be achieved by plasticizers.

## 5.4　COMPOSITIONS BASED ON LIQUID THIOKOL AND BITUMEN

As it was mentioned above, thiokol sealants are widely used in constructions for sealing of interpanel joints, glass packet and airfield seams [1, 2, 4]. They are known to be also used as fluid mastic roof coverings [51, 52]. Prerequisites for using thiokol sealants in constructions are their high resistance to UV radiation and ozone, water resistance and a broad operating temperature range from –60 to +100°C. The advantage of thiokol sealants (as well as of any other oligomer formulations, for example, urethane or silicon blends) is an opportunity of one-step application of coatings of any width and shape and their high vulcanization rate (1 h and more), that allows to use such coating in 24 h after application, as well as these coatings can be applied using various machinery. An important factor here is deficit of thiokol sealants, first of all due to ecological aspects of their production and their cost, which is 3–4 times more, than the cost of the most widespread and durable roof covering materials based on "EPDM" rubber. It sets a problem for application o thiokol sealants for roof covering. Therefore, it is necessary to make thiokol sealants much cheaper for their successful application in construction industry. To achieve these purposes, sealants are usually blended by extra fillers and plasticizers.

A prospective solution here is the use of materials, which are available and widely used in construction and will have the properties fillers and plasticizers if added to thiokol sealants. Bitumen is of undoubted attraction in this aspect, as they are widely used in construction for preparation of blends designed for waterproofing, roof coverings and road construction. Coal tars, mineral rubbers and bitumen are known to be used as modifying additives for thiokol sealants [1, 4, 51–53]. Researches in this area have found a possibility of thiokol sealants modification by bitumen. But a research concerning interrelation of composition and properties of such compositions has not been done yet. The results provided below are intended for solving this problem.

It is necessary to know a chemical structure of bitumen for better understanding of their influence on curing processes in oligomer systems. As phaltenes, being the basic component of bitumen, mainly consist of aromatic poly conjugated compounds and can inhibit radical curing process under certain conditions [54]. Indeed, it was found, that the effectiveness of oligoestera crylates curing in bitumen by organic peroxide depends on curing temperature and reaches maximum in the temperature range of 140–160°C [54–56]. It is explained by the "local activation" effect, found by professor Mejikovscii, S. M., with colleagues. This effect is typical for compounds with poly conjugated bonds [42, 57]. To authors' opinion, poly conjugated as phaltene compounds act as a "trap" for free radicals at a temperature beyond 160°C, which form during decay of organic peroxides. This effect suppresses radical processes. The effectiveness of radical processes is small at 140°C [42]. However, it is necessary to consider a fact of substantial promoting of curing processes by acidic compounds as well as their retarding by basic compounds, when analyzing curing of polysulfide oligomers by metal oxides and dioxides [1, 23]. As petroleum bitumen is produced by air oxidation, it is natural, that they are acidic: 1–3 mg of KOH'g [58]. Bitumen adsorption on the surface of oxidant's particles (in particular, manganese dioxide) inhibits interaction of manganese dioxide with PSO's SH-groups. As PSO curing by metal oxides and dioxides can be described as an ionic-radical process [20, 59], both aforementioned facts must be considered when designing bitumen-thiokol formulations.

### 5.4.1  MODIFICATION OF THIOKOL SEALANTS BY PETROLEUM BITUMEN

There are researches concerning kinetics of bitumen-modified liquid thiokol curing. The method was the variation of $^{13}C$ NMR spin-spin relaxation time $T_2$ [46, 60]. The object was a formulation containing liquid thiokol of grade 2 (the content of SH-groups is 2 weight %), P–803 carbon black (30 mass parts) and BN–70'30 construction bitumen. 50 mass parts of bitumen were introduced into a sealing paste (liquid thiokol and carbon black) at the temperature of +120°C. Manganese dioxide (paste number 9), Sodium bichromate and zinc oxide were used as vulcanizing agents. As Fig. 5.27 demonstrates, formulations with no vulcanizing agents (zero point on the ordinate axis) have lower initial $T_2$ time ($T_2$ of a sealing paste is 4.0 ms) for increased bitumen content (bitumen's $T_2$ is 0.3 ms). We can conclude from Fig. 5.27, a, that the increase of bitumen content at the initial stage of curing by manganese dioxide leads to the decrease of curing rate. The exception is the composition with 5 mass parts of bitumen, whichdemonstrates faster curing, than the reference formulation.

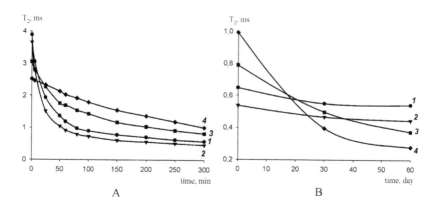

**FIGURE 5.27**   The dependence of $T_2$ on bitumen content (mass parts) during curing of thiokol sealants by manganese dioxide (A – the initial curing period, B – the final curing period): 1 – 0; 2 – 5; 3 – 25; 4 –50.

Further progress of the process is shown in the Fig. 5.27b. It describes sealant's curing after gel-formation process. It can be seen, that the final $T_2$ value of the sealant with 50 mass parts of bitumen reaches $T_2$ value of clear bitumen. It indicates inhibition of liquid thiokol curing by bitumen, which, however, does not impair the degree of curing, but, on the contrary, increases it. If zinc dioxide is used, acceleration of curing processes is proportional to added bitumen (Fig. 5.28). The loss of viability (the start of gel formation) and complete curing states are reached much earlier for zinc dioxide, than for manganese dioxide (Figs. 5.27 and 5.28). However, viability values of thiokol sealants also indicate different influence of bitumen on their curing rate, depending on a vulcanizing agent nature. If manganese dioxide and sodium bichromate are used, addition of up to 25 mass parts of bitumen makes curing 4–5 times slower, while they become 2 times faster if zinc dioxide (Table 5.8).

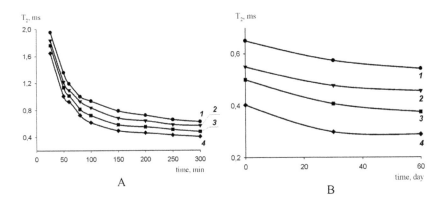

**FIGURE 5.28**    The dependence of $T_2$ on bitumen content (mass parts) during curing of thiokol sealants by zinc dioxide (A – the initial curing period, B – the final curing period): 1 – 0; 2 – 5; 3 – 25; 4 – 50 mass part.

**TABLE 5.8**    Basic Physico-Chemical Characteristics of Thiokol-Bitumen Compositions

| Bitumen, mass part | Parameters | $MnO_2$ | $Na_2Cr_2O_7$ | ZnO |
|---|---|---|---|---|
| 0 | Conventional strength, MPa | 2.25 | 2.12 | 2.8 |
| | Relative elongation % | 100 | 50 | 130 |
| | Viability, hours | 1 | 0.5 | 4 |
| | Adhesion to concrete, MPa | 10 adhes | - | - |
| 5 | Conventional strength, MPa | 2.4 | 2.2 | 2.5 |
| | Relative elongation % | 120 | 60 | 140 |
| | Viability, hours | 1.2 | 0.7 | 3.5 |
| | Adhesion to concrete, MPa | 14 adhes | - | - |
| 10 | Conventional strength, MPa | 1.73 | 2.0 | 2.2 |
| | Relative elongation % | 140 | 80 | 150 |
| | Viability, hours | 2 | 1.5 | 3 |
| | Adhesion to concrete, MPa | 18 adhes | - | - |
| 25 | Conventional strength, MPa | 1.43 | 1.63 | 1.8 |
| | Relative elongation % | 200 | 30 | 170 |
| | Viability, hours | 4 | 2.5 | 2 |
| | Adhesion to concrete, MPa | 16 cohes. | - | - |

**TABLE 5.8**    *(Continued)*

| Bitumen, mass part | Parameters | MnO$_2$ | Na$_2$Cr$_2$O$_7$ | ZnO |
|---|---|---|---|---|
| 50 | Conventional strength, MPa | 1.05 | 1.08 | 1.1 |
| | Relative elongation % | 50 | 10 | 50 |
| | Viability, hours | 8 | 5 | 1 |
| | Adhesion to concrete, MPa | 8 cohes | - | - |

The properties of bitumen modified thiokol sealants are given in Table 5.8. As the tabledemonstrates, the influence of bitumen on properties of sealants depends on the nature of a vulcanizing agent. If manganese dioxide or sodium bichromate is used, the strength parameter reaches maximum at 5 mass parts of added bitumen. Small quantities of bitumen seem to favor more active participation of liquid thiokol in a curing reaction thanks to improved adsorptive and, later, chemical interactions with manganese dioxide. Increased adsorption may be caused by the presence of malthene fraction in bitumen, which is well soluble in thiokol and increases its molecular mobility. It is also confirmed by denser network of transversal bonds, both physical and chemical (Fig 5.29) [46, 60, 61]. It should be noted, that bitumen will unlikely form chemical bonds with liquid thiokol in reaction conditions (20 0C) and considering activity of constituent double bonds [62]. Observed acceleration of curing after addition of 5 mass parts of bitumen may be caused by bitumen's hetero atoms (nitrogen, vanadium, nickel etc.), which are able to catalyze oxidation of liquid thiokol's SH-groups by manganese dioxide [63]. More bitumen additives cause a plasticizing effect, increasing relative elongation, reducing strength and leading to a certain decrease of chemical d.t.b. of sealants, no matter, what vulcanizing agent is used [64] (Table 5.8; Fig. 5.29).

Further increase of bitumen content in composition (above 15–20 mass part increases gradually the physical contribution to t.b.d. it seems to be caused by formation of continuous bitumen phase in sealant, where asphaltene particles can physically interact both with each other and with thiokol and carbon black. It is certainly probable, as phaltenes in bitumen are known to be strengthening additives, which can physically, interact and form a coagulation network. This network exerts influence on the whole set of properties [65, 66]. Introduction of bitumen leads to a substantial increase of sealant's adhesion to concrete independently on the nature of a vulcanizing agent. When bitumen content in a sealant is 25 mass parts and more, its tear from concrete type changes from adhesive to cohesive. This effect is proved by observed decrease of sealants' strength.

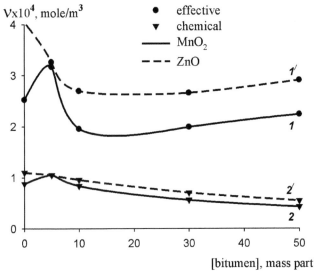

**FIGURE 5.29** The dependence of chemical bonds density (1, 1' – effective, 2, 2'– chemical) on bitumen content in manganese dioxide cured sealant 1, 2 – $MnO_2$; 1' 2'–ZnO.

Bitumens are known to have excellent waterproofing properties and to minimally swell in water [67]. Although thiokol sealants are also high water-resistant, they are not as good as bitumen. Therefore, more bitumen provides manganese dioxide cured sealants with better water resistance (Fig. 5.30). This tendency remains if zinc oxide or sodium bichromate is used. The analysis of bitumen influence on curing kinetics and properties of forming sealants suggests a proposition of ionic-radical process as the most probable mechanism of liquid thiokol curing [20, 59], including adsorption-desorption stage. We can suppose, that bitumen, being a "trap" for radicals, inhibits their formation. In addition, oxidation processes are inhibited due to acidic properties of bitumen and reduced activity of manganese dioxide because of its adsorption on bitumen surface. When liquid thiokol is oxidized by zinc oxide without heating (in normal conditions), curing occurs via formation of mercaptide bonds with their further adsorption on a surface of zinc. A direct proof is the ability of vulcanizates to swell and even dissolute in such polar solvents as benzene and toluene.

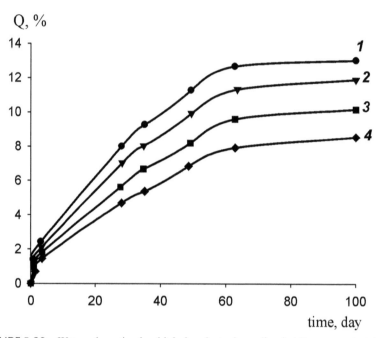

**FIGURE 5.30**   Water adsorption by thiokol sealants depending in bitumen content (mass parts): 1 – 0; 2 – 5; 3 – 25; 4 – 50. (20°C, curing by manganese dioxide).

$$2\ RSH\ +\ ZnO\ \longrightarrow\ \sim\!RSZnSR\!\sim\ +\ H_2O$$

$$\sim\!RSZnSR\!\sim\ +\ ZnO\ \longrightarrow\ \sim\!R\text{-}S\underset{\underset{ZnO}{Zn}}{\diagdown\diagup}S\text{-}R\!\sim$$

$$\sim\!R\text{-}S\text{-}Zn\text{-}S\text{-}R\!\sim\ \begin{cases}\nearrow (t,\,S)\ \sim\!R\text{-}S\text{-}S\text{-}R\!\sim\ +\ ZnS \\ \searrow (t)\ \ \ \ \sim\!R\text{-}S\text{-}R\!\sim\ +\ ZnS\end{cases}$$

When heating is carried out or sulfur is added, desorption processes are possible with further formation of disulfide bonds. As curing of liquid thiokol involves ionic mechanism, and zinc oxide is a weak base, bitumen is an acidic promoter here, as it adsorbs on zinc surface and activates it. Therefore, there are the following conclusions on modification of thiokol sealants by bitumen [46, 60, 61, 64]:

1. Addition of up to 50 mass parts o bitumen increases sealant adhesion to concrete, improve their water resistance and elastic properties, while their strength slightly decreases.
2. Bitumen inhibits curing of thiokol sealants by manganese dioxide due to its inhibiting influence on radical processes and deactivation of manganese dioxide caused by adsorption of bitumen on its surface.
3. Bitumen accelerates curing of liquid thiokol by zinc dioxide that may be reasoned by amphoteric character of zinc dioxide, as it acts as a base in conditions of thiokol curing.

## 5.4.2 MODIFICATION OF PETROLEUM BITUMEN BY LIQUID THIOKOL

A rather prospective way to modify bitumen is their blending with curable oligomers with urethane mane chain or based on liquid Thiokol [42, 52, 68, 69]. Their application will considerably simplify the technology of making compositions, as it shortens the time of oligomer dissolution in bitumen and prevents addition of plasticizing solvents. Thus, there is a principal possibility of making compositions with improved range of properties. If curable oligomers are used, the influence of media and bitumen composition on a curing rate must be considered.

An obligatory prerequisite for successful application of curable oligomeric compositions is their minimal reactivity at synthesis and processing as well as sufficiently high curing rate on products being sealed. Curing agents for oligomers are supposed to promptly and effectively cure them in bitumen forming solid elastomeric matrix with enclosed bitumen in order to exclude possible phase separation in the future. The latter disadvantage is a property of more or less all known "bitum-polymer composition" and it deteriorates their frost resistance and deformability with the course of service life.

One of main conditions for improvement of bitumen properties is its compatibility with polymers used for modification. This property is estimated by the solubility parameter (dp). dp value of liquid thiokol is 18–19.2 $(MJ\dot{\,}m^3)^{0.5}$ [70] and it is close to dp value of aromatic component of malthene fraction in bitumen, which is $18.0(MJ\dot{\,}m^3)^{0.5}$ and of resins, which is $18.6(MJ\dot{\,}m^3)^{0.5}$ [71]. Similarity of dp values makes it possible to predict good compatibility of liquid thiokols with all grades of road bitumen and low-viscous grades of construction and roof bitumen.

Commercial liquid Thiokol of grade 2 with SH-groups content of 2,25% was used for bitumen modification. Thiokol was added to bitumen in a form of a sealing paste, which contained 30 mass parts of P–803 carbon black in addition to Thiokol. Thiokol was cured in bitumen by two oxidative curing agents: $MnO_2$ in a state of a curing paste (paste number 9) with excess factor (molar) to SH-groups content of 1.7; and 10 mass parts of zinc oxide with stearic acid (1 mass part). Diphenylguanidine was used as a vulcanization promoter; its content in the blend was 1% of sealing paste. Thiokol content was varied from 5 to 50%.

Dynamic viscosity was measured on "Polymer RPE–1 MZ" viscometer with "cylinder-cylinder" type measuring cell. Measurements were carried out at temperature of 95°C. Impulse NMR (60 MHz) was used for measuring the duration and degree of thiokol vulcanization in bitumen, as well as to study polymer-bitumen interactions. It was studied, how the process of thiokol vulcanization in bitumen influences spin-lattice ($T_1$) and spin-spin ($T_2$) relaxation times [72].

A curing rate of compositions strongly correlates with the order of components added to bitumen. If thiokol sealant with a curing agent is added to bitumen melt, thiokol will be cured immediately after addition to bitumen and form a separate phase. It has been determined, that a promoter and a curing agent should first be added to bitumen melt and after they distribute homogeneously, thiokol is introduced and a homogeneous mixture will form. Thiokol sealant homogeneously distributes in bulk bitumen in this case and "finds" reactive centers for vulcanization.

As a result, bitumen systems will grow in viscosity, as in other cases of bitumen elastication. Obtained viscosity data made it possible to plot flow curves of bitumen, modified by 40% thiokol compositions depending on holding time (Fig. 5.31). When values of shear rate are low (below lg g = 0,7 s⁻¹) g area can be seen in the curve for initial bitumen, where the flow pattern is closed to Newtonian one and viscosity drops abruptly (lg g> 0,7 s⁻¹) because of destruction of bitumen coagulation structure (non-Newtonian flow pattern) [73–75].

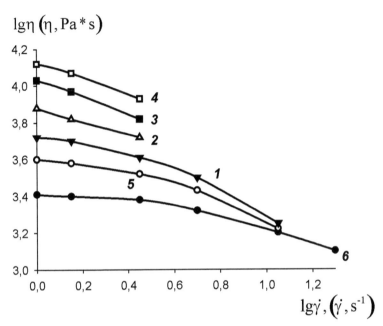

**FIGURE 5.31**    The dependence of viscosity h of BN–70`30 bitumen, modified 40% thiokol composition on the shear rate gradient and a holding period (days): 1 – original bitumen; 2 – 6; 3 – 9; 4 – 32; 5 – bitumen with a curing paste; 6 – bitumen with thiokol, no curing agent.

Modification by thiokol changes flow pattern of bitumen to non-Newtonian in the area of lg g below 0.7 s⁻¹,while the viscosity is higher, than for original bitumen. When composition's holding period is increased, viscosity grows and the flow pattern becomes more deviated from non-Newtonian type [75–77].

Introduction of liquid thiokol (40%) without a curing agent leads to a considerable decrease of bitumen's viscosity, while a flow pattern stays almost unchanged (Fig. 5.31, curve 5). The effect of a cutting paste is similar (4%) (Fig. 5.31, curve 6). Therefore, both uncured thiokol and its curing agent exert a plasticizing influence on malthene and resin fractions of bitumen.

Change of cured thiokol content in bitumen from 5 to 50% (Fig. 5.32), increases viscosity and makes it more dependent on shear rate in its studied range. Bitumen-thiokol systems flow via decomposition of structural formations even at small values of g. The largest increase of viscosity is observed at thiokol concentration in bitumen above 25% (Fig. 5.32, curves 4–6). In this case properties are determined by thiokol itself, which starts to form a continuous network, i.e., phase inversion occurs.

Data in Fig. 5.32 also indicate, that even after holding a composition for 30 days, liquid thiokol is mainly cured via its linear elongation. Increased curing times evidently favor formation of three-dimensional structures (networks) and such compositions cease to flow.

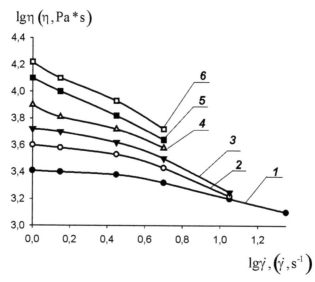

**FIGURE 5.32**   The dependence of BN–70ʹ30 bitumen viscosity on the content of thiokol: 1–0; 2–5; 3–15; 4–25; 5–40; 6–50% (the holding period is 30 days, 23°C).

Therefore, bitumen-thiokol systems are kinetically unstable when cured. Instability increases with thiokol content and a holding period. When thiokol content in a system is 25% and above, more viscous phase forms and delamination in bitumen phase is observed at the critical shear rate. Thiokol (not bitumen) determines viscous properties of a system, i.e., phase inversion occurs, that may lead to system delamination [72, 78].

The change of main performance characteristics of bitumen-thiokol systems on thiokol concentration and formation time (1, 20, 60, 365 days) [77, 78] was studied. However, it should be considered, that a roof can heat up to 80–90°C and vulcanization time of thiokol in bitumen will be considerably shorter.

As Fig. 5.33 demonstrates, that the softening temperature ($T_{soft}$) of bitumen formulations increases both with thiokol content in a composition and the holding period of bitumen-thiokol blends (1, 20, 40, 365 days). Obtained data indicate, that the properties of modified bitumen change dynamically, that is stipulated by curing processes in liquid thiokol. However, curing is slow is not complete even in 40 days. It is caused by a substantial inhibition of end SH-group oxidation in liquid thiokol put into acidic medium.

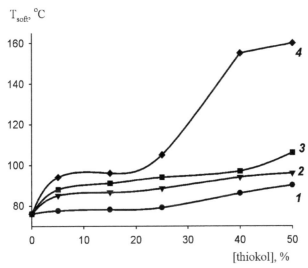

**FIGURE 5.33** The dependence of the softening temperature of bitumen-thiokol compositions on thiokol content and holding period (days): 1–1;2–20:3–40;4–180

It should be noted (Fig. 5.33), that the maximum growth of $T_{soft}$ after complete curing of bitumen-thiokol formulation (365 days) can be observed in two dosage areas. The first area is below 5%, when thiokol is fully compatible with aromatic fraction of malthenes and resins, is fully soluble in them and acts as a plasticizer. The second area is thiokol content above 25%, when a continuous phase of curing thiokol forms. In other words, as it was mentioned above for the study of rheological properties of compositions, the properties of a whole blend are determined here by liquid thiokol properties.

Individual introduction of liquid thiokol, carbon black and a curing paste has been studied to estimate the influence of liquid thiokol curing processes in bitumen and formation of a three-dimensional network on $T_{soft}$ change. $T_{soft}$ behavior has been determined to be almost the same for added sealing paste and thiokol itself in 24 h. Introduction of carbon black causes, as expected, monotonous increase of $T_{soft}$ of bitumen composition from 76°C to 89°C. An unexpected effect is up to 25% (from 76°C to 83.5°C) increase of bitumen's $T_{soft}$, if only thiokol is added. It may be caused by a plasticizing effect of thiokol's dissolution in malthene fraction that favors better ordering of as phaltene particles in bitumen and formation of stronger network of physical bonds. When thiokol content is above 25%, $T_{soft}$ of bitumen composition drops abruptly, that may be reasoned by thiokol separation into an individual phase.

Estimations of penetration of modified bitumen (Fig 5.34) show a substantial influence of thiokol curing processes on properties of compositions. For example, in a day after curing was initiated, a plasticizing effect is observed at thiokol concentration up to 25%, while larger thiokol additives decrease $P_0$ values. There is no plasticizing effect at a high degree of curing (60 days passed) and $P_0$ value decreases, when thiokol content is above 25%. Similar effects are observed at 0°C. $P_0$ decrease caused by introduction into bitumen of up to 25% of liquid thiokol or a sealing paste with ho curing agent may, as above, be stipulated by possible growth of ordering of as phaltene particles in bitumen. When liquid thiokol or sealing paste content is above 25%, P0 value drops abruptly due to a plasticizing effect [76, 78].

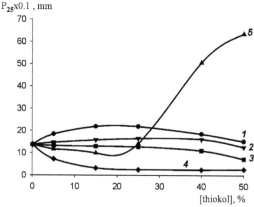

**FIGURE 5.34** The dependence of BN–70`30 bitumen's $P_{25}$ parameter on thiokol content and the duration of curing: 1 –1; 2 –20; 3–40; 4–365 days; 5 – a composition with no curing agent.

Addition of thiokol with a curing agent decreases ductility from 3 to 1 sm (in 7 after curing at 23±2°C). It confirms, that liquid thiokol curing processes occur in bitumen matrix.

The behavior of concentration dependences ductility and penetration indicate a considerable influence of forming continuous thiokol matrix on properties, when the content of curing thiokol is 20–25%, while a three-dimensional network forms at higher degrees of curing. It substantially improves water resistance (Fig. 5.35) and the increase of this parameter is additive to thiokol content in the composition.

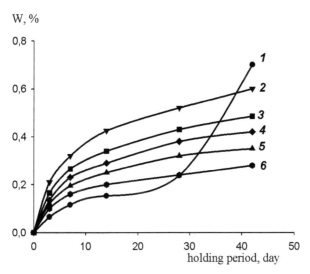

**FIGURE 5.35**   The influence of BN70'30 bitumen's water resistance (W) on thiokol content and the holding period: 1–reference sample; 2–5%; 3–15%; 4–25%; 5–40%; 6–50% of thiokol.

As Table 5.9 demonstrates the increase of thiokol content in bitumen formulations slightly enhances their strength and relative elongation. It is quite logical due to high filling of bitumen by thiokol: a two-phase system forms, which is a polymer network impaired by bitumen inclusions. This effect is also confirmed by rheological research [76].

**TABLE 5.9**   Compositions of bitumen-thiokol blends and their basic properties

| Composition | $T_{soft}$, °C | P x 0.1 mm | | M, °C (R=8 mm), | s, MPa | e, % |
|---|---|---|---|---|---|---|
| | | At 25°C* | At 0°C* | | | |
| BN–70'30 | 76 | 13 | 8 | 0 | - | - |
| +5% of thiokol | 88 | 12.7 | 7.5 | -5 | 0.2 | 105 |
| | 94 | 2.5 | 1.5 | -15 | | |
| +15% of thiokol | 91 | 12.7 | 10.0 | -7 | 0.23 | 140 |
| | 96 | 2.5 | 1.5 | -25 | | |
| +25% of thiokol | 95 | 12.8 | 11.5 | -8 | 0.25 | 150 |
| | 105 | 1.5 | 1.2 | -35 | | |
| +40% of thiokol | 101 | 11.0 | 9.5 | -10 | 0.29 | 170 |
| | 155 | 0 | 0 | -49 | | |
| +50% of thiokol | 105 | 7.0 | 7.0 | -10 | - | - |
| | 160 | 0 | 0 | -55 | | |

*Note: a numerator – 60 days have passed; a denominator – 1 year has passed.

There were estimations of influence of thiokols cured by zinc dioxide (2nd method of curing) on bitumen properties.

Zinc dioxide is much more active curing agent for thiokol in bitumen, than $MnO_2$ at the room temperature already. NMR method has been used to study the influence of thiokol in bitumen vulcanization process on spin-lattice $T_1$ and spin-spin $T_2$ relaxation times (Fig. 5.36) in order to confirm our suggestion on acceleration of thiokol vulcanization by zinc oxide [72, 78].

$T_1$ is determined to be almost insensitive to vulcanization, probably due to a governing role of paramagnetic admixtures (ZnO, $MnO_2$). $T_2$ values certainly decrease in 20–30 min after thiokol vulcanization by ZnO has started. It indicates a distinct slowdown of rotational and translational molecular motion in thiokol itself and/or malthene fraction of bitumen.

Figure 5.37 represents $T_{soft}$ dependence on thiokol sealant concentration in bitumen. $T_{soft}$ has been shown to grow from 76°C to 105°C with the increase of thiokol content in formulations with ZnO, while $T_{soft}$ $MnO_2$

changes from 76°C to 91°C within the same one day. In other words, these data confirm our proposal on ZnO's active curing effect on thiokol in bitumen.

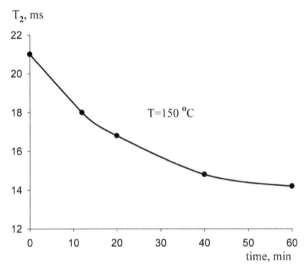

**FIGURE 5.36**    The dependence of $T_2$ of ZnO-cured bitumen-thiokol composition on the duration of curing. (Thiokol content is 40%, curing temperature is 150°C).

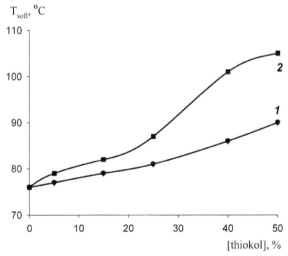

**FIGURE 5.37**    The dependence of $T_{soft}$ of bitumen composition on thiokol concentration. Curing time is 24 h. (1 – curing by $MnO_2$, 2 – by ZnO).

Different activity of $MnO_2$ and ZnO is caused by the influence of media and bitumen composition on liquid thiokol curing processes. Thiokol is completely curing by $MnO_2$ in bitumen in several months only [72, 76, 78]. It correlates well with the information about inhibiting activity of acidic compounds in reactions of thiokol curing by metal dioxides as well as about inhibiting properties of bitumen in relation to radical processes. If ZnO is used, various organic acids, such as levulinic acid, are selected for promotion of a curing process [79]. The rate of curing by zinc oxide in bitumen is typical for thiokol sealants. Indeed, the dependence of penetration ($P_{25}$, $P_0$) of zinc oxide formulations on thiokol sealant content (Fig. 5.38) is determined to decrease from 13 to 8 and from 8 to 6 at 25°C and 0°C correspondingly already in 24 h after the start of curing, indicating high curing rate (Fig. 5.38, curves 1 and 3). Penetration values for compositions with $MnO_2$ (curves 2 and 4) grow. It means, that constituent thiokol has not vulcanized yet and acts as a plasticizer.

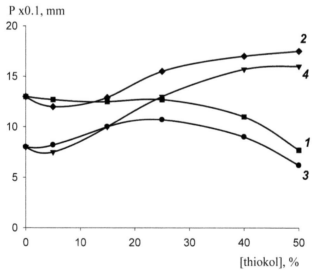

**FIGURE 5.38**    The dependence of P value of bitumen-thiokol compositions on thiokol concentration, when cured by ZnO (curves 1 and 3); or $MnO_2$ (curves 2 and 4); $P_{25}$–1,2; $P_0$–3,4.

Table 5.10 represents values of basic characteristics of bitumen-thiokol systems, when ZnO (numerator) or $MnO_2$ (denominator) are used as vulcanizing agents. As the table shows, formulations with zinc oxide have properties better, than those with $MnO_2$ after the same curing period (one day) elapsed. It may mean that vulcanization by ZnO is faster.

TABLE 5.10   Compositions of Bitumen-Thiokol Blends and Their Basic Properties

| Composition | $T_{soft}$, °C | P, x 0,1 mm | |
|---|---|---|---|
| | | at 25°C* | at 0°C* |
| BN-70`30 | 76 | 13 | 8 |
| +5% of thiokol | 79 | 12.5 | 8.2 |
| | 77 | 11 | 7.5 |
| +15% of thiokol | 84 | 12.5 | 10 |
| | 78 | 12.9 | 10 |
| +25% of thiokol | 89.5 | 12.7 | 10.7 |
| | 81 | 15.5 | 13 |
| +40% of thiokol | 91 | 11 | 9.0 |
| | 86 | 17 | 15.7 |
| +50% of thiokol | 105 | 7.7 | 6.2 |
| | 90 | 17.5 | 15.9 |

*Note: a numerator – formulations, vulcanized by ZnO; a denominator–formulations, vulcanized by $MnO_2$. Properties have been estimated after.

The analysis of liquid thiokol curing in bitumen and properties of related compositions reveals a profound bitumen influence on the degree of curing. Indeed, vulcanization of TG (ТГ) with no bitumen, cured by manganese dioxide, completes in 14 days, while addition of bitumen increases this time up to 365 days. Curing of TG (ТГ) by zinc oxide in a blend with bitumen is rapid and completes in 14 days, while experiments have shown, that this process does not occur at all without bitumen.

Taking into consideration, that the proposed mechanism of liquid thiokol curing by $MnO_2$ is ionic-radical [20], inhibiting effect of bitumen may by mainly caused by poly conjugated aromatic structures in as phaltenes, which are traps for radicals [42, 54]. On the contrary, ZnO curing is accelerated by levulinic or propionic acid [79, 80]. Vulcanization of liquid thiokol in bitumen by ZnO seems to be catalyzed by bitumen. It is also possible; that liquid thiokol curing by ZnO has ionic mechanism and bitumen does not is inhibitor in this case. The reaction of liquid thiokol curing by ZnO in bitumen can be described by the following equation:

$$\sim 2\ RSH + ZnO \longrightarrow \sim R\text{-}S\text{-}Zn\text{-}S\text{-}R\sim + H_2O$$

It should be noted, that mercaptide bonds in the main chain deteriorate water resistance of compositions. The analysis of prepared bitumen systems has revealed their high water resistance. Considering the temperature of bitumen modification (160°C) and the presence of sulfur, that is typical for bitumen in Tatarstan, the reaction providing bitumen with high water resistance may look like this:

$$\sim R\text{-}S\text{-}Zn\text{-}S\text{-}R\sim + S \longrightarrow \sim R\text{-}S\text{-}S\text{-}R\sim + ZnS$$

Therefore, above researches have determined, that liquid thiokol curing by $MnO_2$ is considerably reduced due to inhibiting effect of bitumen. The curing rate of systems with ZnO is much higher, than with $MnO_2$. It influences properties of compositions. There is a suggestion that 25% and more added thiokols, a continuous phase of thiokol sealant forms in bitumen, stipulating all the main properties of modified bitumen.

The analysis of liquid thiokol influence on properties of modified bitumen reveals a distinct effect of bitumen (thiokol) even at small concentration (5 mass parts). The maximum modifying effect on the range of properties is observed for 20–25 mass parts of modifying additive. It can be explained in both cases by formation of a continuous phase: bitumen phase if a thiokol sealant is modified and thiokol phase if bitumen is modified. Emerging of the second continuous phase leads in both cases to improvement of expected properties of the main matrix (the main phase), which are typical for formulations (bitumen, thiokol), forming the second phase. These properties are relative elongation, adhesion and water resistance for thiokol sealants and a considerable broadening of operating temperature range for bitumen.

Thus, research, which have been done in the discussed area, resulted in design of bitumen modified thiokol sealants and bitumen-thiokol compositions prospective for sealing seams on airfields, production of filling mastic roof coverings as well as waterproofing of various building structures [81–83].

## KEYWORDS

- fillers
- influence
- modification
- oligomer
- polysulfide
- properties
- sealants

## REFERENCES

1. Smyslova, R. A., Shvec, V. M., & Sarishvili, I. G. (1991). Application of Curable Sealants in Construction. A Review by VNIINTI and Economy of Industrial Construction Materials (in Russian), Series 6, (2) 30 p.
2. Averko-Antonovich, L. A., Kirpichnikov, P. A., & Smyslova, R. A. (1983). Polysulfide Oligomers & Related Sealants (in Russian) Leningrad: Himija, 128 p.
3. Smyslova, R. A. (1974). Liquid Thiokol Sealants (in Russian).–Moscow: CNIIT Jeneftehim, 83 p.
4. Lucke, H. (1994). Aliphatic Polysulfides Monograph of an Elastomer Publisher Huthig & Wepf Basel, Heidelberg, New York 191 p.
5. Idijatova, A. A., Khakimullin, Yu. N., Vol'fson, S. I., & Liakumovich, A. G. (1997). Abstracts of 4th Russian theoretical & practical conference on rubber "Feedstock & Materials for Rubber Industry. Present State & the Future" (in Russian). Moscow 281–282.
6. Idijatova, A. A., Khakimullin, Y. N., & Vol'fson, S. I. (1999). Zhurnal prikladnoj himii (in Russian) (5) 849–852.
7. Volfson, S. I., Khakimullin, Y. N. (2000). XX Symposium of rheology collection of abstracts. Karacharovo, Russia.p.194
8. Idijatova, A. A. (1999). Technical Sciences Candidate's Dissertation (in Russian). Kazan.
9. Lipatov, Ju. S. (1977). Physical Chemistry of Filled Polymers (in Russian) Moscow Himija 265 p
10. Sellers Dzh., Tunder, F. (1968). Highly Hispersed Strenghtening Silica Acids & Silicates 341–357 in the book "Strenghtening of Elastomers" (in Russian). Ed. by John Krausa. Moscow. Himija. 483p.
11. Idijatova, A. A., Khakimullin, Y. N., & Liakumovich, A. G. (1995). Abstracts of IXth International Young Scientists Conference on Chemistry & Chemical Technology (in Russian) Moscow, 136.

12. Idijatova, A. A., Khakimullin, Yu. N. (1997). Collected papers of theoretical & practical conference "Production & Consumption of Sealants and Other Construction Materials: Present State and Prospectives" (in Russian). Kazan' 63–64
13. Averko-Antonovich, L. A., Kirpichnikov, P. A. (1978). Materials of International Conference on Caoutchouc & Rubber (in Russian) Kiev 44–51.
14. Idijatova A. A., Khakimullin, N., & Liakumovich, A. G. (1996). Abstracts of theoretical & practical conference "Present State & Prospectives of Synthetic Rubbers, Polysulfide Oligomers & Their Derivatives" (in Russian). Kazan 26.
15. Van-Krevelen, D. V. (1976). Properties & Chemical Structure of Polymers (in Russian) Moscow Himija. 416 p.
16. Solomko, V. P. (1980). Filled Crystallizing Polymers (in Russian) Kiev. «Naukova Dumka». 264 p.
17. Muhutdinova, T. Z., Averko-Antonovich, L. A., Kirpichnikov, P. A., & Muhametsalihova, V. G. (1972). KSTU Transactions (in Russian) 50 153–160.
18. Minkin, V. S., Averko-Antonovich, L. A., & Bezrukov, A. V. (1975). Izv. VUZov, "Chemistry & Chemical Technology" series (in Russian) 18(12) 1953–1956.
19. Averko-Antonovich, L. A., Muhutdinov, A. A., Muhutdinova, T. Z., & Hismatullin, R. A. (1977). Izv. VUZov (in Russian) 20(4) 564–567.
20. Minkin, V. S. (1997). NMR in Commercial Polysulfide Oligomers (in Russian) Kazan. "ABAK" 222 p.
21. Khakimullin, Yu. N., Gubajdullin, L. Ju. (2001). Collected papers of theoretical & practical conference "Present State & Prospectives of OJSC Kazanskij zavod SK" (in Russian). Kazan. 128–129.
22. Khakimullin, Ju. N., Hozin, V. G. (2002). Collected papers of international theoretical & practical conference. "Composite Materials, Theory & Practice" (in Russian) Penza 350–351.
23. Khakimullin, Yu. N. (1997). Collected papers of theoretical & practical conference "Production & Consumption of Sealants and Other Construction Materials: Present State and Prospectives" (in Russian). Kazan. 27–39.
24. Valeev, R. R., Khakimullin, Yu. N., Petrov, O. N., Nefed'ev, E. S., & Liakumovich, A. G. (1999). Abstracts of 6th Russian theoretical & practical conference on rubber "Feedstock & Materials for Rubber Industry from Feedstock to Final Products (in Russian)." Moscow 269–270.
25. Valeev, R. R., Idijatova, A. A., Khakimullin, Yu. N., Muhutdinova, T. Z., Vol'fson, S. I., & Liakumovich, A. G. (2000). Vestnik KGTU (in Russian). (1–2). S. 58–63
26. Valeev, R. R., Idijatova, A. A., Khakimullin, Yu. N., Liakumovich, A. G., Petrov, O. V., & Nefed'ev, E. S. (2000). Abstracts of 7th international conference on chemistry & physical chemistry of oligomers "Oligomers 2000" (in Russian). Perm. 242.
27. Valeev, R. R., Idijatova, A. A., Petrov, O. V., Khakimullin, Yu. N., Nefed'ev, E. S., & Liakumovich, A. G. (2001). Kauchuk i rezina (in Russian) (3) 29–32.
28. Valeev, R. R. (2004). Technical Sciences Candidate's Dissertation (in Russian), Kazan.
29. Valeev, R. R., Khakimullin, Yu. N., Muhutdinova, T. Z., Liakumovich, A. G., & Gubajdullin, L. Ju. (2000). Abstracts of 6th Russian theoretical & practical conference on rubber "Feedstock & Materials for Rubber Industry" (in Russian) 284.

30. Petrov, O. V., Nefed'ev, E. S., Khakimullin, Yu. N., Idijatova, A. A., & Chistjakov, V. A. (2000). Zhurnal prikladnoj himii (in Russian).73(3) 501–504.
31. Petrov, O. V. (1999). Chemical Sciences Candidate's Dissertation (in Russian), Kazan.
32. Valeev, R. R., Khakimullin, Yu. N., Idijatova, A. A., Petrov, O. V., Nefed'ev, E. S., & Liakumovich, A. G. (2000). Book of articles of 7th All-Russian conference "Structure & Dynamics of Molecular Systems" (in Russian). Jal'chik. 501–503.
33. Lipatova, T. Je., Shejnina, L. S. (1980). Catalysis & Polymer Formation Reaction Mechanism" (in Russian) Kiev- Naukova Dumka 128–146.
34. Shejnina, L. S., Lipatova, T. Je., Vladimirova, L. Ju., & Vengerovskaja, Sh. G. (1981). Vysokomol. soed. (in Russian). Series A. 23(3) 559–566.
35. Shaboldin, V. P., Demishev, V. N., Ionov, Ju. A., & Akatova, S. P. (1982). Vysokomol. soed (in Russian).24 A(5) 1099–1102.
36. Lipatov, Ju. S. (1991). Basic Principles of Filling of Polymers (in Russian). Moscow Himija, 264 p.
37. Trostjanskaja, E. B., Pojmanov, A. M., & Nosov, E. F. (1973). Vysokomolek. soed. (in Russian). Series A. 15(3) 612–620.
38. Charlesby, A., Morris, J., Montaque, P. (1969). J. Polym. Sci. 16C (8). r. 4505–4513
39. Nufuri, A. D., Lipatova, T. Je. (1974). Physical Chemistry of Polymer Compositions (in Russian) Kiev, Naukova Dumka. P.28–31.
40. Ferri Dzh. (1963). Viscoelastic Properties of Polymers (in Russian). Moscow- IL. 555 p.
41. Berlin, Al. Al., Vol'fson, S. A., Oshmjan, V. G., & Enikolopov, N. S. (1990). Design Principles for Polymer Composites (in Russian) Moscow Himija, 240p.
42. Mezhikovskij, S. M. (1998). Physical Chemistry of Reactive oligomers (in Russian) Moscow. Nauka. 232p.
43. Elchueva, A. D. (1989). Technical Sciences Candidate's Dissertation (in Russian), Kazan.
44. Sharifullin, A. L. (1991). Technical Sciences Candidate's Dissertation (in Russian), Kazan.
45. Kurkin, A. I., Khakimullin, Yu. N., Gubajdullin, L. Ju., & Liakumovich, A. G. (2001). Abstracts of 8th Russian theoretical & practical conference on rubber "Rubber Industry. Feedstock, Materials, Technology" (in Russian). Moscow 347–348.
46. Kurkin, A. I. (2001). Technical Sciences Candidate's Dissertation (in Russian), Kazan,.
47. Suhanov, P. P., Averko-Antonovich, L. A., Elchueva, A. D., & Dzhanbe-kova, L. R. (1996). Kauchuk i rezina (in Russian). (3) 28–32.
48. Habarova, E. V., Saharova, M. A., Harakterova, L. V., & Morozov, Ju. L. (1997). Kauchuk i rezina (in Russian) (3) 17–21.
49. Rajt, P., Kamming, A. (1973). Polyurethane Elastomers (in Russian). Leningrad Himija, 304 p.
50. Hozin, V. G. (1997). Collected papers of theoretical & practical conference "Production & Consumption of Sealants and Other Construction Compositions" (in Russian), Kazan, 9–20.
51. Shul'zhenko, Ju. P., Grigor'eva, L. K. (1993). Polymer Roofing & Waterproofing Materials. Analytical Survey (in Russian). Moscow, VNIIJeSM, Issue 236 p.

52. Nuralov, A. R., Breeva, G. I., Gulimov, A. G. (1977). Stroitel'nye materialy (in Russian), (9) 8.
53. Author's Certificate 382658 SSSR, MKI S 08 N 13°00, Mastika Bitizol.
54. Gorelov, Ju. P., Vitel's, L. Je., Mezhikovskij, S. M. (1988). Zhurnal prikladnoj himii (in Russian) 61(3) 522–526.
55. Merkin, A. P., Gadzhily, R. G., Vitel's, L. Je., & Mezhikovskij, S. M. (1984). Azerb. him.zhurnal (in Russian), 4 117–121.
56. Nadzharjan, S. N. (1991). Technical Sciences Candidate's Dissertation (in Russian). Institute of Chemical Physics, AN SSSR. 436.
57. Berlin, A. A., et al. (1972). Chemistry of Polyconjugated Systems (in Russian) Moscow Himija. 271 p.
58. Kisina, A. M., Kucenko, V. I. (1983). Polymer-bitumen Roofing & Waterproofing Materials (in Russian). Leningrad: Strojizdat, 134 p.
59. Coates. R. G., Gilbert, B. C., Lee, T. P. C. (1992). I. Chem.Soc.Perkin Trans. 2, 1387
60. Kurkin, A. I., Khakimullin, Yu. N., & Liakumovich, A. G. (2002). Kauchuk i rezina (in Russian) (6), 18–20.
61. Kurkin, A. I., Khakimullin, Yu. N., Petrov, O. V., Nefed'ev, O. S., & Liakumovich, A. G. (2000). Abstracts of 7th Russian theoretical & practical conference on rubber (in Russian), Moscow, 291–292.
62. Posadov, I. A. et al. (1984). Abstracts of All-Russian conference (in Russian), Tbilisi Moscow, 21–24.
63. Petrochemistry (in Russian). Ed. by. Sjunjaeva, Z. I., L. (1984), 263–313.
64. Khakimullin, Yu. N., Murafa, A. V., Hozin, V. G., & Kurkin, A. I., et al. (2000). Abstracts of 7th international conference on chemistry & physical chemistry of oligomers "Oligomers" (in Russian), Moscow-Perm-Chernogolovka, (2000), 294.
65. Bodan, A. N., Kostjuk, B. L. (1987). Asphaltene & Tarry Compounds as Ingredients for Rubber Blend. A Topical Survey (in Russian). Moscow, CNIITJe- neftehim, 68 p.
66. Rozental,' D. A., Posadov, I. A., Popov, O. G., & Pauku, A. N. (1981). Fraction of Heavy-oil Residues: Methods of Detection & Calculation of Structural Parameters (in Russian), Leningrad, LTI im. Lensoveta, 80 p.
66. Gorshenina, G. I., Mihajlov, N. V. (1967). Polymer-bitumen Composite Materials (in Russian), Moscow, Nedra, 240 p.
67. Khakimullin, Yu. N., Murafa, A. V., Sungatova, Z., & Khozin, V. G. (1998). International conference on Polymer characterization Texas USA 62
68. Kac, B. N., Glotova, N. A. (1980). Collected Papers of VNII strojpolimer (in Russian), (53), "Polymer Construction Materials" 78–95.
69. Muhutdinova, T. Z., Averko-Antonovich, L. A., Prokudina, K. N., & Kirpichnikov, P. A. (1973). KSTU Transactions, (50), 141–145.
70. Rozental', D.A., Tabolina, L. S., & Fedosova, V. A. (1988). Modification of Bitumen Properties by Polymer Additives (in Russian) Moscow, Surveys in Oil-processing & Petrochemical Industry, "Oil Processing" series 6 49p.
71. Khakimullin, Yu. N., Murafa, A. V., Nagumanova, Je. I., Sungatova, Z. O., Hakimov, A. M., Sundukov, V. I., & Hozin, V. G. (1999). Book of articles of All-Russian conference "Structure & Dynamics of Molecular Systems" (in Russian). Jal'chik 212–221.

72. Nagumanova, Je. I., Murafa, A. V., Sungatova, Z. O., Khakimullin, Yu. N., & Hozin, V. G. (1998). Book of articles of 7th All-Russian conference "Structure & Dynamics of Molecular Systems" (in Russian) Jal'chik. 243–245.
73. Khakimullin, Y. N., Sungatova, Z. O., Kurkin, A. J., Nagumanova, E. J., Murafa, A. V., Khozin, V. G., & Zaikov, G. E. (2002). Journal of the Balkan Tribological Association 8(1) 1–6.
74. Khakimullin, Yu. N., Sungatova, Z. O., Kurkin, A. I., Nagumanova, E. I., Murafa, A. V., Khozin, V. G., & Zaikov, G. E. (2001). Russian Polymer News 6(2) R. 29–32.
75. Sungatova, Z. O. (1999). Technical Sciences Candidate's Dissertation (in Russian). Kazan.
76. Khakimullin, Yu. N., Sungatova, Z. O., Kurkin, A. I., Nagumanova, Je. I., Murafa, A. V., Hozin, V. G., Zaikov, V. G., & Zaikov, G. E. (2000). Plasticheskie massy (in Russian) (1) 36–38.
77. Khakimullin, Yu. N., Murafa, A. V., Sungatova, Z. O., Nagumanova, Je. I., & Hozin, V. G. (2000). Mehanika kompozitnyh materialov (in Russian). 36(5) 701–710.
78. Patent 4165426 USA, MKI S08 G 75´04.
79. Application for European patent 0061242, Inventions in USSR and abroad (in Russian). (1983). 18–58, 30.
80. Sungatova, Z. O., Muruzina, E. V., Murafa, A. V., Khakimullin, Yu. N., & Hozin, V. G. (1999). Abstracts of 5th International Conference on Intensification of Petrochemical Processes "Neftehimija 99" (in Russian). Nizhnekamsk, 114–115.
81. Khakimullin, Y. N., & Hozin, V. G. (1999). Abstracts of international conference "Chemistry and Technology of Composite Materials, Based on Bitumen Emulsions and Modified Bitumen (in Russian). Minsk. 58.
82. Patent, RF 2179986, MPK C08 L 95´00.

**CHAPTER 6**

# TECHNOLOGICAL AND SERVICE PROPERTIES OF SEALANTS, BASED ON POLYSULFIDE OLIGOMERS

## CONTENTS

## 6.1   THE INFLUENCE OF COMPOSITION OF THIOKOL SEALANTS ON THEIR STABILITY DURING PRE-USE STORAGE

Polysulfide oligomer sealants (PSO) are highly gas-proof. Their designed set of properties remains after curing even if curing agent dosage deviates largely from its optimal value. Sealant components are satisfactory stable during pre-use storage. The curing rate of PSO sealants is highly sensitive to air humidity on the one hand, while final properties of sealant are fully insusceptible to it on the other hand.

Aforementioned advantages of liquid thiokol sealants stipulate their application in aircraft and mechanical engineering, in a form of three-component compositions, as well as in construction, in a form of two-component compositions, where precise weighting of components and observance of temperature and humidity requirements are not always possible, when sealant is being prepared or applied.

Pre-use stability of sealant's components underlies both technological and service properties of related sealants. Therefore, it is interesting to study the influence of compositions of sealant's components (sealing and curing pastes) on their stability during storage.

Thiokol sealants usually combine two or three components [1–3]:
  (1) a sealing paste, which includes polysulfide oligomer, filler and adhesive additive;
  (2) a vulcanizing paste, with combines inorganic oxidant (manganese or lead dioxides or sodium bichromate), a plasticizer, a curing rate regulator and thixotropic additives;
  (3) curing promoter, such as diphenylguanidine (DPG) or thiuram, which is usually added to a vulcanizing paste for construction sealants or is used as the third component of tailored sealants [1–3]?

Sealing pastes usually contain epoxy resin in addition to liquid thiokol and filler (carbon black, titanium dioxide or chalk). The storage period of such pastes depends strongly on their composition.

Sealing pastes of different industrial batches have been determined [4] to be sometimes different in their curing activity. We suggested that it could be caused by duration and conditions of their storage.

That is why it has been studied how sealing paste's composition (liquid thiokol type, addition of epoxy resin and filler type) influences is stability [5]. The reference properties were viscosity, varied with time, HS-groups content and spin-spin relaxation time ($T_2$) values. The content of SH-groups was determined by iodometry (TU 38.50309–93).

The research was done on NVB–2-grade industrial liquid thiokols (viscosity at 25°C–8.8 Pa·s, SH-groups content–3.45% weight), first-grade thiokols ( the viscosity at 25°C 22.5 Pa·s, SH-groups content–2.67% weight) and E-40 epoxy Diane resin (epoxy groups content–15.0% weight).

Mixing of liquid thiokol with P-803-grade (ТУ П–803) carbon black has been determined to move a flow curve from its Newtonian type with more added filler. Filled blends behave as plastic liquids: they obey New-tonian law in areas, where stresses overcome a fluidity limit (Fig. 6.1). Storage of liquid thiokol mixtures with P–803 carbon black has been found to slightly reduce a molecular mobility (measured by $T_2$), while HS-groups content remains virtually unchanged. It fully correlate with the ear-lier proved [6] fact of adsorptive interaction between mixture components (carbon black and liquid thiokol).

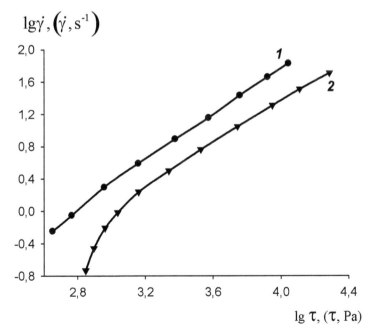

**FIGURE 6.1** The flow curves of liquid thiokol (1) and its mixture with 30 mass parts of P-803 carbon black (2) at 25°C. (g–shear rate, s⁻¹; t–stress, Pa).

Mixing of liquid thiokol with E-40 epoxy resin results in decrease of spin-spin relaxation time, especially for high oligoepoxide dosages in U-30 E-10 paste (Fig. 6.2a). At the same time, the content of SH-groups reduces (Fig. 6.2b). These processes are active during first two months of storage. Liquid thiokol viscosity is inessential here, as obtained dependences for I-grade and NVB-2-grade thiokol look similar. Samples with 5 mass parts of E-40 resin (curves 3 and 4) remain stable after one month of storage, but they contain only 65–70% of initial quantity of HS-groups.

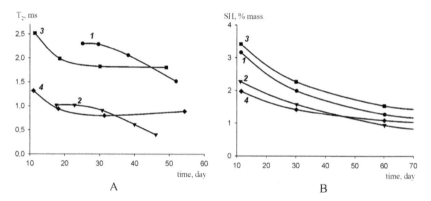

A                 B

**FIGURE 6.2** Changing of spin-spin relaxation time $T_2$ (A) and the content of HS-groups (A) during storage of liquid thiokol mixtures of I (2, 4) and II (1,3) grades in pastes with 5 mass parts (3, 4) and 10 mass parts (1, 2) of E–40 epoxy resin.

Pastes with epoxy resin are unstable because end groups of liquid thiokol and epoxy oligomers interact:

$$\sim RSH + H_2C\!-\!\overset{H}{\underset{O}{C}}\!-\!R \longrightarrow \sim R\!-\!S\!-\!\overset{H2}{C}\!-\!\overset{H}{\underset{OH}{C}}\!-\!R\sim$$

Occurrence of this reaction is confirmed by differences in NMR $^1$H spectra (Fig. 6.3.) of liquid thiokol mixtures with E-40 and individual oligomer spectra (Table 6.1).

**TABLE 6.1**   Chemical shift of $^1$H nucleuses (ppm) for E-40 Epoxy Resin, Liquid Thiokol and Their Mixture with the Mass Ratio of Components 1 : 2

| Oligomer type | $C_6H_4$ | $CH_3$ | $OCH_2O$ | $SCH_2$ | $H_2C-CH-$ $O$ | $OCH_2$ |
|---|---|---|---|---|---|---|
| E-40 | 7.2 | 1.6 | - | - | 2.75 | - |
| PCO | - | - | 4.5 | 2.8 | - | 4.5 |
| E-40÷PCO | 7.2 | 1.6 | 4.7 | 2.9 | 2.4 | 4.7 |

$$H_2C-CH-$$
$$O$$

A typical feature of a mixture of oligomers is decreased NMR signal intensity for group, increased intensities of $SCH_2$ and $OCH_2$ group signals, as well as shifting of $SCH_2$, $OCH_2$ and $OCH_2O$ groups' signals to the area of weaker fields.

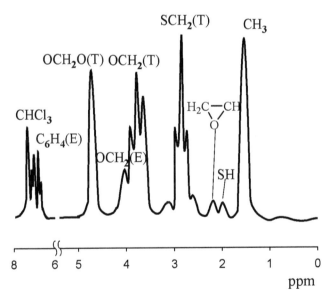

**FIGURE 6.3**   High-resolution NMR spectra for liquid thiokol (T), epoxy resin (E) and their mixture with the mass ratio of components 2:1.

Therefore, it has been found, that oligomer and filler do not interact in U-30 paste and such mixture can be stored for a long time (more, than a year) assuming no air oxygen. U-30 E–5 and U-30 E–10 sealing pastes are unstable at storage. Interaction of oligomers in the first paste completes in the first storage month, while increased dosage of E–40 resin makes this period 1–2 months longer. In any case, pastes with epoxy resin demonstrate decrease of SH-groups concentration distinctly growing with time, while these groups are necessary for further vulcanization of thiokol sealants. Taking this fact into consideration, we recommend storing liquid thiokol and E–40 resin separately and mixing them later, when making a sealant.

| The content of E-40 resin, mass parts, per 100 mass parts of liquid thiokol | 5 | 10 | 5 | 10 |
|---|---|---|---|---|
| Liquid thiokol grade | NVB-2 | NVB-2 | Ist grade | Ist grade |
| $T_2$ values, ms: | | | | |
| immediately after mixing | 2.43 | 1.90 | 1.79 | 1.10 |
| one month of storage | 1.60 | 1.27 | 0.80 | 0.71 |
| two months of storage | 1.56 | 0.55 | 0.68 | 0.50 |
| The content of HS-groups, %: | | | | |
| immediately after mixing | 3.43 | 3.45 | 2.67 | 2.67 |
| one month of storage | 0.70 | 0.64 | 1.4 | 0.87 |
| two months of storage | 0.61 | 0.36 | 1.26 | 0.67 |

Prolongation of storage period of liquid thiokol-titanium dioxide mixture with no influence of air oxygen was found to reduce spin-spin relaxation time $T_2$, (Fig. 6.4). In addition, these changes occur independently on the viscosity of selected oligomer. If there are two oligomers in a sealing paste: polysulfide oligomer and E-40 resin, than these changes are more definite:

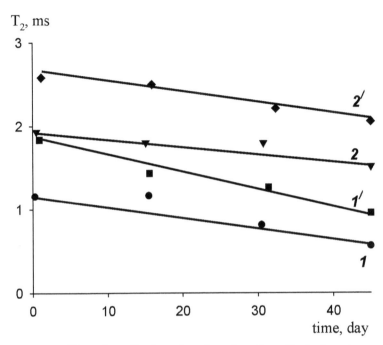

**FIGURE 6.4**   Changing of spin-spin relaxation time $T_2$ during storage of the following compositions (in mass part): 1–II grade liquid thiokol (100) and $TiO_2$ (80); 1'–the same with E-40 resin (10); 2 I grade liquid thiokol (100) и $TiO_2$ (80); 2'– the same with E-40 resin (10).

Titanium dioxide is known [7] to be hydrophilic and able to interact with various compounds forming sufficiently strong bonds, which do not break in flow processes. We thought that titanium dioxides behave similar to zinc salts, which enter into complexation with sulfur-containing compounds [8], and can interact with disulfide bonds of polysulfide oligomer:

$$\sim R\text{-}S\text{-}S\text{-}R\sim \xrightarrow{TiO_2} \begin{matrix} \sim RS^- \\ \\ \sim RS^- \end{matrix} TiO_2 \longrightarrow \sim RS^{\bullet} + \sim RSTiO_2$$

Donor-acceptor interactions with $TiO_2$ or ordinary adsorption of oligomer molecules on filler's surface will themselves lead to inhibition of spin-spin interaction (Fig. 6.4). However, if complexation with further generation of radicals takes phase, PSO can be oxidized in chain reaction and

EPR spectral line shape for oligomer-titanium dioxide mixture will change at first. Mixtures were heated up to 40 and 70°C to make possible formation of radical easier and EPR spectra were taken after these mixtures were exposed to these temperatures for 1–2 weeks.

Obtained data have shown that the shape of EPR spectral line and g-factor are not changed in studied conditions. Therefore, we can suggest that no chemical interactions accompany mixing of polysulfide oligomer with titanium dioxide and that observed decrease of molecular mobility is caused by adsorption of oligomer's segments on the surface of filler particles.

Therefore, the epoxy resin is not supposed to be introduced into a composition of thiokol-based sealing pastes, as it interacts with polysulfide oligomer. The more E-40 resin is added to a sealing blend, the more changed $T_2$ values and HS-groups content become. In general, all commercial sealants with epoxy resin (U 30 MES5, U 30 MES10, UT 32, etc) lose considerably their technological properties during pre-use storage. It eventually deteriorates service properties [2.9–13]. Therefore, thiokol and epoxy resin are recommended to be stored separately. Epoxy resin is desirable to be added to a sealing composition either in a form of copolymer with thiokol or as thiokol adducts.

## 6.2   THE INFLUENCE OF POLYSULFIDE OLIGOMER NATURE ON TECHNOLOGICAL AND SERVICE PROPERTIES OF SEALANTS

Sealants containing so-called "thiol-containing polyesters" as polysulfide oligomers (PSO) are successfully used in construction in the last decades. They are such PSOs, as, ZL–616, PM, permapol P–2 and TPM-2 polymer. All the listed PSOs are based on polyethers, usually polyoxipropyleneglycols, and contain 1–2% weight of sulfur in a form of sulfhydryl end groups. Such sealants exceed thiokol products in process ability, adhesion to various substrates and maximum permissible deformation, they are highly air and frost-resistant. However, they stay behind liquid thiokol sealants in strength, oil and gasoline resistance and water resistance [14–18].

There was research on improvement of sealants based on liquid thiokol or TPM-2 polymer by their combination [19–22]. A prerequisite for obtaining sealants with advanced properties via mixing of such PSOs is the presence of end SH-groups and Therefore, an opportunity of using a common curing agent. However, their compatibility should be considered as well as their adsorption activity to fillers and difference in curing rates [19, 23, 24]. The research objects were liquid thiokol formulations (the content of SH-groups = 3.1% weight, viscosity 11.5 Pa•s) and TPM-2 polymer blends (the content of SH-groups = 2.43% weight, viscosity 2.0 Pa•s), filled with 80 mass parts of natural chalk (MTD-2). Curing was carried out by a paste-like manganese dioxide. The ratio between sealing and curing pastes was 2:1. Formulation's curing rate was estimated by the viability parameter and $T_2$ value. Sealants had been cured for 14 days after the loss of viability before they were tested. The temperature was 23±2°C. The ratio of TPM-2 polymer and liquid in sealants was varied in the range of 0–100 mass parts correspondingly. Changing of properties was analyzed during consecutive substitution of TPM-2 polymer to liquid thiokol.

The analysis of flow curves of chalk-filled oligomers has revealed three interesting facts (Fig. 6.5):

(1) TPM-2–polymer formulations are less viscous at the same degree of filling and Therefore, are easier to process.

(2) Formulations with prevailing TPM-2 polymer are more thixotropic, than thiokol blends. It is explained by earlier revealed [25] ability of TPM-2 polymer to participate in formation of three-dimensional network of labile hydrogen bonds.

(3) A concentration dependence of viscosity has S–type shape (Fig. 6.5).

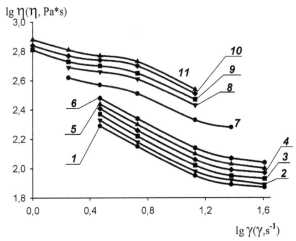

**FIGURE 6.5** The dependence of dynamic viscosity on the shear rate of PSO formulations (TPM-2 polymer: liquid thiokol): 1–100:0; 2–95:5; 3–90:10; 4–85:15; 5–70:30; 6–50:50; 7–30:70; 8–15:85; 9–10:90; 10–5:95; 11–0:100

When the content of TPM-2 polymer is predominant (≥50% weight), the growth of viscosity is insignificant. A flow pattern seems to be stipulated by TPM-2 polymer's viscous properties in this case. Further increase of TPM-2 polymer in sealant and corresponding increase of thiokol content lead to phase inversion and stepwise growth of viscosity. A degree of compatibility of TPM-2 polymer and liquid thiokol seems to be taken into consideration when explaining properties of their combinations. A parameter, which characterizes compatibility of polymers with high degree of probability, is a solubility parameter (d). When d values of compared polymer are equal, these polymers are usually thermodynamically compatible and they are incompatible, when these values are different. The solubility parameter of TPM-2 polymer, calculated by the method of [26], is 17.6 (MJ/m³), while its value for liquid thiokol is from 18.0 to 19.2 MJ/m³ [27, 28].

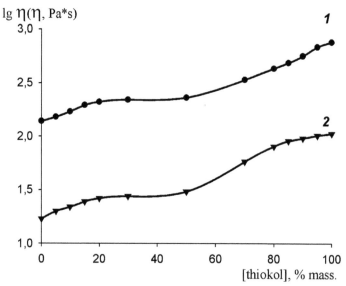

**FIGURE 6.6**   The concentration dependence of filled (1) and unfilled (2) formulations on PSO content.

Taking into consideration, that TPM-2 polymer and used commercial liquid thiokol, based on di-b(chloroethyl)formal are thermodynamically incompatible, fully mutually insoluble and easily separate to individual layers id left still after mixing, we can propose, that TPM-2 polymer is a dispersion medium, when the content of thiokol in a sealant is below 50% weight. When the ratio of oligomers is 1:1, phase inversion occurs.

There was research on the influence of liquid thiokol on the viability and curing kinetics of TPM-2 polymer. Viability estimations made for studied compositions (Fig. 6.7) indicate, that TPM-2 polymer is less active in reactions of oxidation by manganese dioxide, than liquid thiokol. Also has been found to be less active in reaction involving manganese dioxide for ZL-616 [14]. A more active curing system is required for preparation of sealants, whose viability will be as high as liquid thiokol formulations. Similar result can be achieved if liquid thiokol is added to TPM-2 polymer formulations. Indeed, addition of more than 5 mass parts of thiokol leads to abrupt acceleration of a curing process. When more thiokol is added, curing accelerates too, but to a lesser extent.

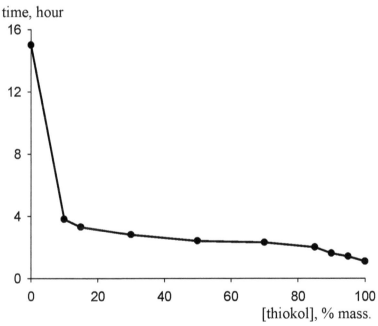

**FIGURE 6.7** The dependence of sealants viability on thiokol content.

As it has been noted above, there is a good deal of subjectivity in the method of viability estimation, which is usually used for cold-setting sealants (GOST 25621–83 ) and its accuracy is ±30 for formulations with viability values within 2–24 h. A more precise estimation of oligomer's formulation on its curing rate has been achieved with [13]C NMR. As figures demonstrate, $T_2$ values (this parameter is used for measuring a curing rate via molecular mobility alterations) decrease, when thiokol is added to a sealant. Abrupt $T_2$ decrease(indicating the increase of vulcanization rate) seems to be caused by structural differences of oligomers and the presence of impurities in liquid thiokol, which are oxidation catalysts. Considering, that small thiokol additives (5–10 mass parts) do not form the continuous phase of thiokol in its blends with TPM-2 polymer, acceleration of a curing rate may be caused by thiokol sulfur, which is a catalyst for TPM-2 polymer curing. A curing rate varies additively, when TPM-2 polymer is consecutively changed by thiokol. It reaches the maximum value, when thiokol content is 80%. When the content of thiokol is from 80 to 100%, the rate remains constant. Thus, curing kinetics seems to be fully stipulated by thiokol's activity, when its content is 80% and above. The curing rate value is higher for filled compositions independently on PSO type and the ratio of oligomers (Fig. 6.8). Addition of chalk accelerates curing of TPM-2 polymer formulations more, than other ones, as it was mentioned above (Chapter 5).

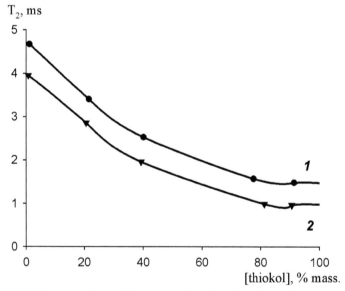

**FIGURE 6.8**   The dependence of spin-spin relaxation time $T_2$ on the content of liquid thiokol and the duration of vulcanization: 1– in 100 min; 2– in 1000 min.

As we expected, a partial substitution of TPM-2 polymer by thiokol favors increase of sealant's strength (Fig. 6.9). Three distinct areas can be found on the strength curve of sealants with varied PSO composition:

1.  Introduction of up to 30% weight of thiokol leads to slight increase of strength. Sealant's strength is stipulated in this area by the strength of continuous 3D-network of TPM-2 polymer macromolecules.
2.  When thiokol content is from 30 to 80% weight, the increase of strength is observable. A homo polymeric 3D-network of thiokol molecules seems to form in this concentration range together with copolymer structures. It exerts deeper influence on sealant's strength with more added thiokol.
3.  Addition of more, than 80% weight off thiokol results in rapid increase of strength. Its value depends now on thiokol's vulcanization structures only.

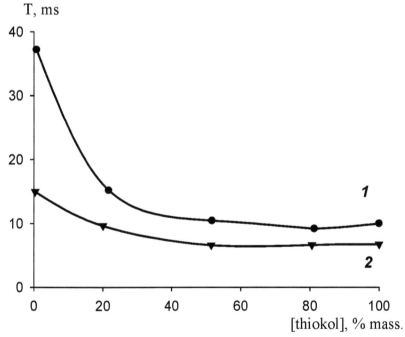

**FIGURE 6.9**   The dependence of vulcanization time (meaning that T value falls down the times) on liquid thiokol's content in sealant: 1 – no chalk; 2 – 100 mass parts of chalk.

When TPM-2 is partially substituted by liquid thiokol (up to 15–20%), relative elongation value does not change (Fig. 6.10). More added thiokol causes the loss of elastic properties, which has minimum at TPM : thiokol ratio from 3:7 to 1.5:8.5 and then increases.

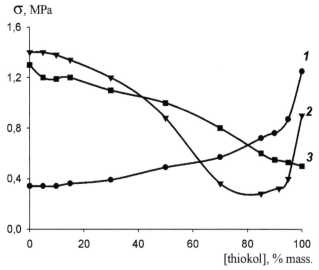

**FIGURE 6.10**   The dependence of sealant properties on thiokol content: 1–conventional strength; 2–relative elongation; 3–adhesion to duralumin.

An extreme of relative elongation function is explained by ambiguous behavior of a curing process involving formation of both homo polymer and copolymer networks. Therefore, deformation properties deteriorate in the middle area of oligomer ratios.

The behavior of sealant's adhesion to duralumin (Fig. 6.10) may be explained by the same factors. When TPM-2 polymer content in sealant is no less, than 50%, its structures are responsible for adhesion to duralumin and this parameter endures a slight change. When thiokol content is sealant is above 50%, a loss of adhesion is observed.

Degrees of swelling in water and toluene have similar behavior (Fig. 6.11).

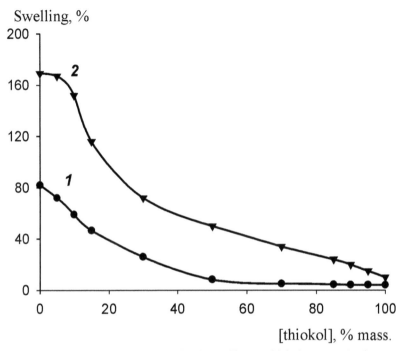

**FIGURE 6.11**    The dependence of sealant's swelling on thiokol content: 1 – in water (7 days); 2 – in toluene (14 days).

TPM-2 polymer sealant is not waterproof enough due to lack of hydrophilic fragments in the main chain and Therefore, is not recommended for sealing of products being constantly exposed to water. Addition of thiokol increases water resistance, and when its content is above 50%, water resistance reaches typical values of thiokol sealants.

The analysis of variation of strength and adhesion properties and swelling of sealants in water and toluene reveals, that properties of a sealant are fully determined by properties of the main polymer, so it is TPM-2 polymer, in the case it is substituted by up to 50% weight of thiokol. When the portion of thiokol increases, sealant's properties become closer to the properties of thiokol sealants. It should be noted, that if copolymer structures form, properties would vary additively, while most parameters change rapidly, when the ratio of oligomers is 1:1.

In connection with foresaid, we can suggest creation of homopolymer structures by thiokol and TPM-2 polymer as interpenetrating networks during curing of sealants containing both TPM-2 polymer and liquid thiokol. Two reasons can explain this effect: 1. Thermodynamic incompatibility of liquid thiokol with TPM-2 polymer; 2. various oxidation activities when cured by manganese dioxide. Indeed, viability and curing rate data (Figs. 6.8 and 6.9) indicate, that liquid thiokol has SH-groups 1.5 times more active, than those of TPM-2 polymer during entire curing process. However, as a common vulcanizing agent is used for both polymers (manganese dioxide), it is logically not to expect 100% selectivity with formation of interpenetrating polymer networks (IPN) only. Therefore, both IPNs and copolymer structures form and there are no abrupt changes of properties with variation of the ratio of oligomers (typical for true IPNs) [20].

Thus, we have studied curing of sealants made of PSO blends of a various nature, as well as their viscous and other properties. Changing of viscous properties, strength, adhesion and swelling in water and toluene has been determined not to change by the additive law. We have shown, that sealants with practically valuable set of properties can be prepared via blending of studied oligomers.

Sealants based on liquid thiokol and TPM-2 polymer enjoy wide application as external sealants for "closed type" interpanel seams in house building [16, 29–36].

A price of PSO makes the largest contribution to a price of a sealant. The world experience of PSOs application shows, that the part of PSO in mastic should be no less, than 30–35% [3, 16, 29]. It primarily relates to liquid thiokol formulations. When thiokol content in sealant is reduced below 30%, strength and adhesion properties deteriorate considerably as well as reproducibility of properties of various batches. However, blending thiol-containing polyester with TPM-2 polymer is prospective for design of sealant, containing 20% of oligomer (Table 6.2) [23, 37].

**TABLE 6.2** The Dependence of Sealant Properties on PSO Content (applied into sample seams)

| Property | PSO content, % | | | |
|---|---|---|---|---|
| | **50** | **30** | **20** | **15** |
| Sealantbased on "TSD" Thiokol | | | | |
| s, MPa | 0.96 | 0.54 | 0.21 | 0.12 |
| $e_{rel}$, % | 220 | 185 | 140 | 75 |
| Break type | cohesive | cohesive | cohesive-adhesive | adhesive |

**TABLE 6.2**    *(Continued)*

| Property | PSO content, % | | | |
|---|---|---|---|---|
| | **50** | **30** | **20** | **15** |
| Sealantbasedon TPM-2 polymer | | | | |
| s, MPa | 0.64 | 0.41 | 0.33 | 0.21 |
| $e_{rel}$, % | 225 | 235 | 180 | 150 |
| Break type | cohesive | cohesive | cohesive | cohesive |

It is explained by the ability of such oligomers to be blended by large quantities of fillers and plasticizers. Even when oligomer content is as low as 15%, related sealants follow GOST 25621–83 requirements (titled "Polymer materials and products for sealing and packing used in construction").

However, considering the quality of raw materials and oligomer itself as well as potential of mixing equipment in Russia, at least 25% of TPM-2 polymer is required to make sealants with highly reproducible properties. The necessary prerequisites for success of such sealants in construction are their technological and service properties in addition to properties, guaranteed by their formulation. The most important technological factors are the viscosity and thixotropicity of blends. A sealant is supposed to be viscous enough for hand filling into a sealing gun in conditions of a construction site, as well as to be easily applied into and remain thixotropic.

Figure 6.12 demonstrates, that formulations with TPM-2 polymer are considerably less viscous, than formulations with liquid thiokol. In addition, when aerosils are added to TPM-2 polymers, compositions with high thixotropicity and good processability can be prepared, in contrast to liquid thiokol blends [25, 38]. Liquid thiokol sealants suit construction site conditions at the temperature as low as –5°C. The advantage of TPM-2 polymer sealants is that their viscosity is less dependent from temperature, andTherefore, they can be used almost year-round at the temperature as low as –20°C. It is confirmed by the analysis of LT–1 [15] and SG–1 [23, 37] grade sealants, which are based on TPM-2 polymer (Fig. 6.12). The viscosity of SG–1 grade sealant grows insignificantly, when temperature reduces from +20°C to–10°C, while the viscosity of AM-05-grade sealants grows many times in the same conditions.

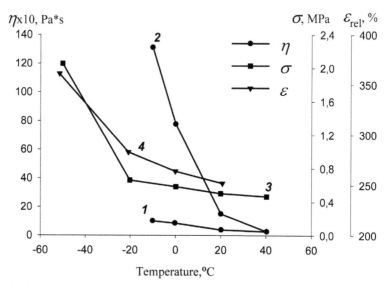

**FIGURE 6.12** The dependence of sealant properties on temperature: 1 – the viscosity of SG-1-grade sealant; 2 – the viscosity of AM-05-grade sealant; 3 – the strength (s) of SG-1; 4 – the relative elongation ($e_{rel}$) of SG-1.

One of the main parameters, characterizing service properties of "seam" sealants, is the value of actual deformation. The maximum actual deformation of liquid thiokol sealants is 25% [29]. According to information from (MNIITEP), the maximum actual deformation of SG-1-grade sealant is 50%. This sealant has high strength properties and relative elongation at temperatures from +40°C to –50°C (Fig. 6.12). As Table 6.3 demonstrates, properties of SG-1-grade sealant remain almost unchanged after aging on air or in water. This sealant is resistant to repeating cyclic deformations with seam's strain amplitude up to 50%.

**TABLE 6.3** Changing of Properties ofSGГ-1-Grade Sealant Exposed to Aging

| Tests | | s, MPa | $e_{rel}$, % | Break type |
|---|---|---|---|---|
| Source data | | 0.49 | 248 | cohesive |
| Ageing | Air (10 days, 70°C) | 0.41 | 207 | cohesive |
| | Water (10 days. 20°C) | 0.31 | 473 | cohesive |

**TABLE 6.3** *(Continued)*

| Tests | | s, MPa | $e_{rel}$, % | Break type |
|---|---|---|---|---|
| 1000 cycles of deformation with the amplitude | I = 25% | 0.49 | 268 | cohesive |
| | I = 50% | 0.41 | 309 | cohesive |
| Express method by "MNIITEP" (ISO) | | 0.31 | 360 | cohesive |

Sealants do not lose their properties after complex exposure (ultraviolet, thermal aging, sprinkling, and chill proofing at –40°C with further cyclic deformations) in express mode.

One of considerable disadvantages of epoxy resin blended thiokol sealants is both instability of sealing paste's viscosity (Fig. 6.13) and of sealant's physico-chemical and adhesive properties depending on pre-use storage duration.

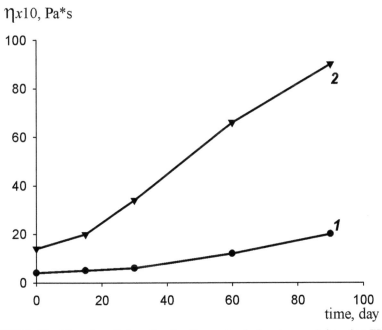

**FIGURE 6.13** Changing of viscosity of sealing pastes during storage: 1–based on TPM-2 polymer; 2–based on liquid thiokol.

It sets a limit for the shelf life of a sealing paste up to 3 months and is mainly caused by ER-liquid thiokol interaction in storage conditions [5, 13].

The viscosity of a sealing paste based on TPM-2 polymer (and therefore, the main properties of a sealant), changes much slower during storage, that's why such compositions have the shelf life of more, than 3 months and their properties are more reproducible, than for liquid thiokol sealants.

Liquid thiokol production technologies, both in Russia and abroad, generates a large amount of wastes (from 2.5 to 5.0 tons of salts and up to 60 $m^3$ of waste waters per one ton of produced thiokol). Several times less salts generate during TPM–2 polymer production, while waste waters do not emerge at all. Therefore, application of liquid thiokol is justified only when its outstanding properties (oil- and gasoline resistance, gas impermeability) reveal in sealants to the best.

Thus, considering requirements to curable sealants, based on PSO, as well aspresent trends, we can see a distinct differentiation in the functional application of sealants. The best products for interpanel seam sealants are ecofriendly and widely available polyoxypropylenemercaptans (TPM-2 polymer), taking technological and service properties, as well as economy into consideration. Liquid thiokol, however, is in deficit and is produced by environment-unfriendly technology. It is used solely in production of glass packets and as a basis for oil-resistant sealants for aerospace industry and mechanical engineering [16, 39–41].

## KEYWORDS

- composition
- oligomers
- polysulfide
- properties
- sealants
- stability
- technological

# REFERENCES

1. Averko-Antonovich, L. A., Kirpichnikov, P. A., & Smyslova, R. A. (1983). Polysulfide Oligomers & Related Sealants (in Russian). Leningrad: Himija, 128 p.
2. Smyslova, R. A., & Kotljarova, S. V. (1976). Handbook on Rubber Sealing Materials (in Russian) Moscow: Himija. 72 p.
3. Lucke, H. (1994). Aliphatic Polysulfides. Monograph of an Elastomer Publisher Huthig & Wepf Basel, Heidelberg, New York. 191 p.
4. Minkin, V. S. (1997). NMR in Commercial Polysulfide Oligomers (in Russian). Kazan. ABAK 222 p.
5. Nefed'ev, E. S., Hakimullin, Ju. N., Polikarpov, A. P., & Averko-Antonovich, L. A. (1986). Izvestija VUZov (in Russian) .29(1), 97–100.
6. Averko-Antonovich, L. A., Muhutdinov, A. A., Muhutdinova, T. Z., & Hismatullin, R. A. (1977). Izvestija VUZov (in Russian), 20(4), 564–567.
7. Kruglickij, N. N., et al, (1978). Kolloidnyj zhurnal (in Russian), 40(4), 676–681.
8. Shershnev, V. A., Hodzhaeva, I. D., Tarasova, Z. N. (1983). Kauchuk i rezina (in Russian) (6) 16–19.
9. Smyslova, R. A. (1974). Liquid Thiokol Sealants (in Russian). Moscow: CNIIT Jeneftehim. 83 p.
10. Minkin, V. S., Deberdeev, R. Ja., Paljutin, F. M., & Hakimullin, Ju. N. (2004). Commercial Polysulfide Oligomers: Synthesis, Vulcanization, Modifcation (in Russian), Kazan', ZAO "Novoe znanie," 176 p.
11. Labutin, L. P. (1982). Anticorrosive & Sealing Materials Based on Synthetic Rubbers. Leningrad: Himija. 213 p.
12. Mudrov, O. A., Savchenko, I. M., & Shitov, V. S. (1982). Handbook on Elastomer Coatings & Sealants in Shipbuilding (in Russian). Leningrad: Sudostroenie 184 p.
13. Muhutdinova, T. Z., Shahmaeva, A. K., Gabdrahmanov, F. G., & Sattarova, V. M. (1980). Kauchuk i rezina (in Russian), 1, 12–15.
14. Smyslova, R. A. (1984). Liquid Thiokol Sealants (in Russian). Moscow: CNIIT Jeneftehim 67.
15. Serebrennikova, N. D., Somova, L. A., Chernyh, A. H., & Sinajskij, A. G. (1986). Stroitel'nye materials (in Russian), 1, 18–19.
16. Khakimullin, Yu. N. (1997). Collected papers of theoretical & practical conference "Production and Consumption of Sealants and Other Construction Materials: Present State and Prospectives (in Russian). Kazan. 27–39.
17. Hakam Singh. (1987). Rubber World, 196(5), 32, 34–36
18. Patent 3923738 USA, MKI C08 G 18/04.
19. Idiyatova, A. A., Khakimullin, Y. N., Volfson, S. J., & Liakumovich, A. G. (1999). Sixth European Symposium on Polymer Blends. Mainz, Germany. 95.
20. Idijatova, A. A., Khakimullin, Yu. N., & Liakumovich, A. G. (2001). Kauchuk i rezina (in Russian). 3 S. 27–29
21. Idijatova, A. A., Khakimullin, Yu. N., & Liakumovich, A. G. (1997). Abstracts of 6th International Conference on Chemistry & Physical Chemistry of Oligomers (in Russian). Kazan 242.

22. Idijatova, A. A., & Khakimullin, Yu. N. (1999). Abstracts of 6th Russian theoretical & practical conference on rubber "Feedstock & Materials for Rubber Industry from Feedstock to Final Products" (in Russian). Moscow. 271.
23. Idijatova, A. A. (1999). Technical Sciences Candidate's Dissertation (in Russian), Kazan, KSTU.
24. Petrov, O. V., Nefed'ev, E. S., Khakimullin, Yu. N., Idijatova, A. A., & Chistjakov, V. A. (2000). Zhurnal prikladnoj himii (in Russian).73(3) 501–504.
25. Idijatova, A. A., Khakimullin, Yu. N., & Vol'fson S.I. (1999). Zhurnal prikladnoj himii (in Russian). (5), 849–852.
26. Van-Krevelen, D. V. (1976). Properties & Chemical Structure of Polymers (in Russian). Moscow Himija 416 p.
27. Muhutdinova, T. Z., Averko-Antonovich, L. A., Prokudina, K. N., & Kirpichnikov, P. A. (1973). KSTU Transactions (in Russian). (50) 141–145.
28. Shvarc, A. G., & Dinzburg, B. N. (1972). Blending of Rubbers with Plastics & Synthetic Resins (in Russian). Moscow. Himija 224 p.
29. Smyslova, R. A., Shvec, V. M., & Sarishvilli, I. G. (1991). Application of Curable Sealants in construction Techniques. A Review VNIINTIJePSM (in Russian), Series 6, (2) 30 p.
30. Burenin, V. V. (2000). Stroitel'nye materialy (in Russian). (11) 16–18.
31. Idijatova, A. A., Khakimullin, Yu. N., & Liakumovich, A. G. (1995). Abstracts of IXth Young Scientists Conference on Chemistry & Chemical Technology "MKHT-95" (in Russian). Moscow. 136.
32. Idijatova, A. A., Khakimullin, Yu. N., & Liakumovich, A. G. (1998). Abstracts of 5th Russian theoretical & practical conference on rubber "Feedstock & Materials for Rubber Industry. Present State & the Future" (in Russian) Moscow 465–466.
33. Khakimullin, Yu. N., Idijatova, A. A., Valeev, R. R., Gubajdullin, L. Ju., & Liakumovich, A. G. (2000). Materials of 21st international annual theoretical & practical conference "Composite Materials in Industry" (in Russian). Jalta- Slavpolikom 144.
34. Valeev, R. R., Idijatova, A. A., Khakimullin, Yu. N., Gubajdullin, L. Ju., & Liakumovich, A. G. (2001). Collected papers of theoretical & practical conference. "Present State & Prospective of OJSC Kazanskij zavod SK" (in Russian) Kazan. 97–102.
35. Gubajdullin, L. Ju., & Khakimullin, Yu. N. (1999). Construction, Architecture & Housing and Communal Services (in Russian) (1) 19–21.
36. Minkin, V. S., Nistratov, A. V., Vaniev, M. A., Khakimullin, Yu. N., Deberdeev, R. Ja., & Novakov, I. A. (2006). Synthesis, Structure & Properties of Polysulfide Oligomers. A survey in collected papers "Izvestija Volgogradskogo Gosudarstvennogo Tehnicheskogo Universiteta" series "Chemistry and Technology of Elementorganic Monomers and Polymer Materials" (in Russian). Volgograd, Issue 3, (1), 9–20.
37. Idijatova, A. A., Khakimullin, Yu. N., & Liakumovich, A. G. (1998). Kauchuk i rezina (in Russian). (5). 30–33.
38. Volfson, S. I., & Khakimullin, Yu. N. (2000). XX Symposium of Rheology. Collection of abstracts. Karacharovo, Russia. 194.
39. Khakimullin, Yu. N., Paljutin, F. M., & Hozin, V. G. (2005). Stroitel'nye materially (in Russian), 10, 69–73.

40. Khakimullin, Yu. N., & Gubajdullin, L. Ju. (2001). Collected papers of theoretical & practical conference "Present State and Prospectives of OJSC Kazanskij zavod SK" (in Russian). Kazan. 128–129.
41. Khakimullin, Yu. N., & Hozin, V. G. (2002). Collected papers of international theoretical and practical conference "Composite Materials, Theory and Practice" (in Russian). Penza. 350–351.

# INDEX

Milton Keynes UK
Ingram Content Group UK Ltd.
UKHW050257161024
449569UK00042B/1757